Springer Theses

Recognizing Outstanding Ph.D. Research

Aims and Scope

The series "Springer Theses" brings together a selection of the very best Ph.D. theses from around the world and across the physical sciences. Nominated and endorsed by two recognized specialists, each published volume has been selected for its scientific excellence and the high impact of its contents for the pertinent field of research. For greater accessibility to non-specialists, the published versions include an extended introduction, as well as a foreword by the student's supervisor explaining the special relevance of the work for the field. As a whole, the series will provide a valuable resource both for newcomers to the research fields described, and for other scientists seeking detailed background information on special questions. Finally, it provides an accredited documentation of the valuable contributions made by today's younger generation of scientists.

Theses may be nominated for publication in this series by heads of department at internationally leading universities or institutes and should fulfill all of the following criteria

- They must be written in good English.
- The topic should fall within the confines of Chemistry, Physics, Earth Sciences, Engineering and related interdisciplinary fields such as Materials, Nanoscience, Chemical Engineering, Complex Systems and Biophysics.
- The work reported in the thesis must represent a significant scientific advance.
- If the thesis includes previously published material, permission to reproduce this must be gained from the respective copyright holder (a maximum 30% of the thesis should be a verbatim reproduction from the author's previous publications).
- They must have been examined and passed during the 12 months prior to nomination.
- Each thesis should include a foreword by the supervisor outlining the significance of its content.
- The theses should have a clearly defined structure including an introduction accessible to new PhD students and scientists not expert in the relevant field.

Indexed by zbMATH.

Zhehao Zhu

Influence of Fine Particles on the Liquefaction Properties of a Reference Sand

Application to the Seismic Response of a Sand Column on a Vibrating Table

Doctoral Thesis accepted by
École des Ponts ParisTech, Marne-la-Vallée, France

Springer

Author
Dr. Zhehao Zhu
School of Civil Engineering
Shanghai Normal University
Shanghai, China

Supervisor
Dr. Jean-Claude Dupla
Laboratoire Navier/Cermes
École des Ponts ParisTech
Marne-la-Vallée, Paris, France

ISSN 2190-5053 ISSN 2190-5061 (electronic)
Springer Theses
ISBN 978-3-031-24298-4 ISBN 978-3-031-24299-1 (eBook)
https://doi.org/10.1007/978-3-031-24299-1

This Springer imprint is published by the registered company Springer Nature Switzerland AG
The registered company address is: Gewerbestrasse 11, 6330 Cham, Switzerland

Supervisor's Foreword

The threat of sand liquefaction to human activities from earthquakes is sufficient to require their attention to the design of civil engineering structures. According to field surveys, the liquefaction of sand containing fine particles has received increased attention over the last thirty years.

The Ph.D. work carried out by Zhehao contributes to properly explaining some contrasting results reported in the literature concerning the influence of fine particles on liquefaction. To achieve this goal, Zhehao first succeeded in establishing a new experimental protocol with high repeatability. An extensive database of reliable monotonic and cyclic triaxial results was developed to emphasize different roles of fine particles (plastic and non-plastic) with two reconstitution methods. Based on these results, a reasonable mechanism was proposed to conclude that the influence of fines highly relies on their relative position to coarse sand grains. Zhehao also participated in developing a physical model setup on the shaking table of the CEA (French Alternative Energies and Atomic Energy Commission). The preliminary tests provided a solid basis and experience for future research. Finally, Zhehao performed numerical simulations with a new calibration method based on artificial intelligence. The clear benefit of such a method is to bypass the traditional trial-and-error procedure involving many human factors.

This Ph.D. work was involved in the project ANR "ISOLATE" (characterizing and Improving SOiLs AgainsT liquEfaction), financed by French National Research Agency (2017–2022), piloted by Prof. Semblat (ENSTA Paris). The principal objective is to establish a reliable framework for liquefaction assessment. The project includes academic partners (Gustave Eiffel University, CentraleSupélec, Centrale Nantes) and industrial partners (CEA, EDF, Solétanche Bachy).

I have been impressed by the maturity of Zhehao and his independence. He managed to stay focused and performed highly complex experiments and numerical simulations. Zhehao is very open minded and communicative. It was a great pleasure to supervise his excellent work, and I am proud to announce that he received the

"2021 Chinese Government Award for Outstanding Self-financed Students Abroad". I am convinced that he is a very dedicated and brilliant person with a strong potential for an academic career at Shanghai Normal University.

Paris, France
November 2022

Jean-Claude Dupla

Parts of this thesis have been published in the following journal articles:

Zhu, Z., Zhang, F., Peng, Q., Dupla, J.C., Canou, J., Cumunel, G., Foerster, E. (2021). Assessment of the loading waveform on the cyclic liquefaction resistance with Hostun 31 sand, Soil Dynamics and Earthquake Engineering, 150, 106919

Zhu, Z., Zhang, F., Peng, Q., Dupla, J. C., Canou, J., Cumunel, G., & Foerster, E. (2021). Effect of the loading frequency on the sand liquefaction behaviour in cyclic triaxial tests. Soil Dynamics and Earthquake Engineering, 147, 106779

Zhu, Z., Zhang, F., Peng, Q., Chabot, B., Dupla, J. C., Canou, J., Cumunel, G., & Foerster, E. (2021). Development of an auto compensation system in cyclic triaxial apparatus for liquefaction analysis. Soil Dynamics and Earthquake Engineering, 144, 106707

Zhu, Z., Zhang, F., Dupla, J. C., Canou, J., Foerster, E., & Peng, Q. (2021). Assessment of tamping-based specimen preparation methods on static liquefaction of loose silty sand. Soil Dynamics and Earthquake Engineering, 143, 106592

Zhu, Z., Zhang, F., Dupla, J. C., Canou, J., & Foerster, E. (2020). Investigation on the undrained shear strength of loose sand with added materials at various mean diameter ratios. Soil Dynamics and Earthquake Engineering, 137, 106276

Zhu, Z., Dupla, J. C., Canou, J., & Foerster, E. (2022). Experimental study of liquefaction resistance: effect of non-plastic silt content on sand matrix. European Journal of Environmental and Civil Engineering, 26(7), 2671–2689

Acknowledgements

First and foremost, I would like to thank my mom. I know well you have left me for many years already. To this world, I might be merely one person, important or perhaps not, but to you I am the whole world. During this journey without you, life was sad at times for accomplishing this Ph.D. work. No matter how long the rain lasts, I believe there will be a rainbow in the end. No matter how sad I may be, I remember how much you love me.

I would like to thank my supervisor Dr. Jean-Claude Dupla, for his extensive scientific supervision. I appreciate, from the bottom of my heart, his kindness, his permanent availability, and the encouragements that he gave me. My profound thanks then go to Dr. Jean Canou for guiding this thesis. His experience and skills were of great importance to help me quickly find answers. I also thank Dr. Gwendal Cumunel for our inspiring discussions and especially for the time he devoted to the development of a laminar shear box. I'd like to thank Dr. Evelyne Foerster, Former Director of the Mechanical and Thermal Studies Department at CEA Paris-Saclay, for all her advice related to the development of two advanced soil containers in shaking table tests.

I want to express my sincere thanks to all the jury members, starting with Prof. Jian-Fu Shao at Polytech Lille, who did me the honour of presiding over my Ph.D. defence. My thanks then go to Prof. Pierre Breul at Polytech Clermont and Prof. Pierre-Yves Hicher at Centrale Nantes, who were kind enough to examine my work thoroughly. I also give my thanks to Dr. Philippe Reiffsteck, Research Director at Gustave Eiffel University, Dr. Emmanuel Javelaud, Researcher at EDF, and Dr. Vincent Crozet, Researcher at CEA Paris-Saclay for their great interest devoted to this work.

The organization of the research work and the achieved results would not have been possible without the priceless helps from my colleagues at the Geotechnical team of the Navier laboratory piloted by Prof. Yu-Jun Cui. Their availability and untiring energies in keeping the edge of experimental apparatus are indispensable for the success of this dissertation. Prof. Jun Yang at the University of Hong Kong is gratefully acknowledged for his support in preparing Fig. 1.31 of the present work. The financial supports provided by the National Natural Science Foundation of China (Grant Nos. 41931286 and 41672348) are deeply acknowledged.

Finally, I owe a special debt to my family and Dr. Qiutong Li who constantly supported me during these years. Words cannot express how grateful I am to have you in my life.

Shanghai, China
November 2022

Zhehao Zhu

Contents

Abbreviations

a	Ground acceleration
b	Active fraction of fine particles
B	Skempthon's B value
CSL	Critical state line
CSR	Cyclic stress ratio
C_g	Social parameter
C_p	Cognitive parameter
C_u	Uniformity coefficient
d	Dilatancy law
d_{50}	Mean grain size (fine particles)
D	Depth
D_{50}	Mean grain size (coarse sand)
D_{50}/d_{50}	Mean diameter ratio
D_r	Relative density
DP	Dry pluviation
DT	Dry tamping
e	Global void ratio
e_c	Critical void ratio
e^{eq}	Equivalent void ratio
e_f	Interfine void ratio
e_{max}	Maximum void ratio
e_{min}	Minimum void ratio
e_s	Skeleton void ratio
f	Loading frequency
f_c	Fines content (divided by soil mass)
f_{thre}	Transition fines content
F	Force
F_c	Fines content (divided by sand mass)
g	Gravitational acceleration
G	Shear modulus
GA	Genetic algorithm

G_s	Specific gravity
h	Height
HHT	Hilber-Hughes-Taylor method
I_A	Arias intensity
I_{Dmat}	Density index of sand matrix
I_p	Plasticity index
I_{pth}	Threshold plasticity index
k	Iteration number
K	Bulk modulus
K_c	Consolidation stress ratio
L_I	Loading intensity
LSB	Laminar shear box
m	Mass
M	Critical stress ratio
MT	Moist tamping
N_{cyc}	Number of cycles
N_L	Number of cycles to liquefaction
NP	Non-plastic
p_{at}	Atmospheric pressure
PGA	Peak ground acceleration
PSO	Particle swarm optimization
PTS	Phase transformation state
p'	Mean effective stress
q	Deviator stress
q_{max}	Peak deviator stress
r^2	Coefficient of determination
r_u	Excess pore water pressure ratio
R	Shape characteristic
SA	Acceleration spectra
SEM	Scanning electron microscope
S_r	Saturation degree
S_u	Undrained shear strength
t	Time
T	Period
T_d	Duration of ground motion
TFC	Transitional fines content
u	Pore water pressure
u_0	Initial pore water pressure (back pressure)
W	Weight factor
WN	White noise
α	Newmark/HHT integrator
β	Newmark integrator
ε_a	Axial strain
ε_d	Deviator strain
ε_v	Volumetric strain

ϕ	Internal friction angle
ϕ^{+}_{inst}	Internal friction angle at the instability state (triaxial compression)
Δt	Time step
Δu	Excess pore water pressure
ΔV	Volumetric variation
Δp_{conf}	Confining pressure increment
Δz	Element size
η	Stress ratio
η^{+}_{car}	Stress ratio at the phase transformation state (triaxial compression)
η^{-}_{car}	Stress ratio at the phase transformation state (triaxial extension)
η^{+}_{inst}	Stress ratio at the initiation of instability (triaxial compression)
η^{-}_{inst}	Stress ratio at the initiation of instability (triaxial extension)
ρ_{w}	Water density
σ'_{a}	Axial effective stress
σ'_{c}	Isotropic consolidation stress
σ'_{d}	Imposed deviator stress
σ'_{r}	Radial effective stress
τ	Shear strain
τ_{f}	Vane shear resistance
ω	Inertia weight
ψ	State parameter

Introduction

In the geotechnical literature, a great variety of damages associated with ground failures have been reported since the foundation soil might develop a flow-type behaviour due to the rapid accumulation of excess pore water pressure, especially under rapid seismic loading. In such a case, the shallow saturated soil can experience a complete loss of shear strength and even behave as a "liquid". This phenomenon is widely known as the generic name "liquefaction". As a result of soil liquefaction, spectacular engineering failures can take place. For instance, the 1964 Niigata earthquake with a magnitude of 7.5 is an iconic example, during which more than 300 local buildings were damaged. Recently, Christchurch in New Zealand was severely hit by a series of seismic events in 2010 and 2011. A variety of destructions were observed, including residential buildings associated with large permanent ground deformation and other underground lifelines.

Over the past sixty years, numerous efforts have been devoted to this field. In laboratory research, the triaxial test has been commonly adopted as a meaningful approach to examining the development of sand liquefaction. As a result, a sound understanding of sand liquefaction has been already acquired. The earlier research mainly focused on clean sand specimens and did not cover the possible influence of fine particles. However, field observations continually suggest that the *in situ* liquefied soil is generally composed of a predominant sand matrix polluted by some fine particles and more than half of liquefied data fell within a small range of fines contents. Unfortunately, the influence of fine particles has not been figured out yet. In the published literature, no common agreements and contradictory results can be found as to whether the introduction of fine particles provides a positive or negative contribution to the global force chain. In this context, the main objective of this thesis is to clarify and quantify the influence of fine particles involved in the host sand matrix on the initiation and the subsequent development of sand liquefaction. Provided that saturated sandy elements containing fewer fine particles are more likely to liquefy, only low fines contents have been involved in the experimental programme and the density index of sand matrix has been taken as the controlled parameter while preparing test specimens to conserve the prevailing sand matrix. Both monotonic and cyclic triaxial tests have been adopted. In order to underline the influence of fine particles associated with the different initial soil fabrics created by different reconstitution methods, two tamping-based methods (moist and dry tamping) were used for the test specimens containing non-plastic fines. Microstructural observations were then performed with the aid of an optical microscope with the purpose of directly visualizing the soil array to differentiate the mechanical role of fine particles at different locations. Based on these observations, a mechanism behind regarding whether the host sand skeleton is broken by the addition of fine particles has been proposed to summarize the influence of fine particles. Cyclic triaxial tests were carried out to verify the above argument and establish the analogies existing between monotonic and cyclic behaviour. In addition, two series of cyclic triaxial tests were conducted on clean sand specimens. One of the features of these two series of tests is the coverage of a wide range of loading frequencies and shapes to emphasize their inherent impacts on the development of sand liquefaction.

The understanding of sand liquefaction acquired from the conventional triaxial tests is only for elementary size, far from representing a true semi-infinite soil extent. 1 g shaking table tests are a helpful approach to get closer to in situ conditions and to evaluate possible scale effects and testing protocols on the soil response (e.g. cyclic vs. dynamic loading). In addition, it can allow to investigate mitigation measures against liquefaction. That was among the challenges addressed by the ISOLATE project, funded by the French National Research Agency (ANR). In our work, two model soil containers have been developed. The first one was sized quite small to check the reconstitution and the saturation methods, as well as the characterization ability of certain conventional geotechnical tools (e.g. vane shear test). The second soil container has been manufactured in large size to explore the sand liquefaction during base shaking in an almost actual engineering size. In particular, flexible foams were attached to the rigid side walls in order to greatly reduce the parasite boundary effects (e.g. wave reflections). To accurately represent the sand liquefaction in a numerical manner, an automatic calibration programme based on an artificial intelligence method (particle swarm optimization, PSO) was programmed in the Python environment. The Dafalias bounding surface model has been taken as an example to demonstrate the main idea. The optimized model parameters were then used to simulate a large-sized shaking table test with the assessment of the possible boundary effect during base shaking.

The present thesis is organized according to the following five chapters. Chapter 1 presents the main achievements of sand liquefaction already reported in the literature. Chapter 2 presents the tested materials (sand and fine particles), equipment and protocols used for triaxial testing. In addition, special attention is paid to introducing a novel auto compensation system, ensuring a correct isotropic consolidation state prior to the shearing stage. Chapter 3 presents the experimental results of the specimens subjected to monotonic loading and the corresponding microstructural observations. Similarly, Chap. 4 first presents the experimental results of the sand-fines specimens subjected to cyclic loading and then investigates the impact of loading frequency and shape on the liquefaction response. Chapter 5 first summarizes some well-known shaking table setups and the corresponding experimental results reported in the literature. Then, the equipment developed in this work is presented with three preliminary tests using the small soil container to validate the reconstitution and saturation method and the entire experimental environment. The Dafalias model is briefly introduced with its key ingredients, and an automatic calibration methodology based on the PSO method is presented. The versatility of the proposed method was carefully assessed by using experimental data of two reference sands (HN31 sand and Ottawa sand F65). Combined with the simulation results of a shaking table test, a cost-effective improvement method for reducing the parasite boundary effect is introduced for the large-sized soil container.

Chapter 1
Bibliographic Review

Check for updates

1.1 History of Earthquake Liquefaction

Derived from the Latin verb "liquefacere", liquefaction is the term nowadays employed in materials science to depict a process that generates a liquid from a granular solid. In the field of soil mechanics, liquefaction is a physical phenomenon in which a saturated or partially submerged soil loses its strength in response to an applied external loading, which is frequently triggered under natural states by the intense shearing induced by earthquakes (Ishihara and Yoshimine 1992).

Around the world, the phenomenon of soil liquefaction owing to earthquakes has been broadly reported throughout human history. More than a half-century has passed since 1964 when two destructive earthquakes occurred in the United States and in Japan that initiated liquefaction research: the Alaska and the Niigata earthquakes. The 1964 Niigata earthquake with a magnitude of 7.5 is an iconic example of soil liquefaction (Fig. 1.1a), which resulted in dramatic damages: (i) approximately 310 local buildings constructed of reinforced concrete were damaged, 200 of which were tilted; (ii) liquefaction-induced piles deformation was possibly the major reason to explain the collapse of the Showa Bridge that was designed based on the most advanced earthquake-resistance engineering at that time and was completed shortly before the earthquake (Yoshida et al. 2007). The 1964 Alaska earthquake with a magnitude of 8.6, also referred to as "Great Alaskan earthquake" or "Good Friday earthquake", was the most powerful earthquake recorded in North American history, during which 131 people were believed to have died. Soil liquefaction and landslide (Fig. 1.1b) were two major causes of structural damages in several communities: 9.6 million m^3 of soils were taken away and the ground was consequently devastated by these huge displacements. In a more recent study, widespread liquefaction was proved responsible for extensive damage to lifelines and residential properties (Cubrinovski et al. 2010) in various parts of Christchurch city during the 2010 Darfield earthquake (New Zealand).

Besides level ground, soil liquefaction risk is also frequently presented in other important geotechnical structures (Lade and Pradel 1990; Lade 1992, 1993) such as

Z. Zhu, *Influence of Fine Particles on the Liquefaction Properties of a Reference Sand*, Springer Theses, https://doi.org/10.1007/978-3-031-24299-1_1

Fig. 1.1 Photographic views of liquefaction disasters: **a** Niigata earthquake; **b** Alaska earthquake (Niigata earthquake from https://en.wikipedia.org/wiki/Soil_liquefaction and Alaska earthquake from https://en.wikipedia.org/wiki/1964_Alaska_earthquake)

embankment dams. One of the best-known examples in the literature concerns the failure of the Lower San Fernando dam (Castro et al. 1992) following the California 1971 earthquake. The experimental studies performed in the field have revealed that (i) the rupture suddenly occurred with the development of liquefaction in a sand layer approximately 20–30 s after the occurrence of the earthquake; (ii) almost all sandy layers generated a flow-type behaviour owing to the rapid build-up of excess pore water pressure. More recently, during the Iwate-Miyagi Nairiku earthquake (2008) with a magnitude of 7.2, a Japanese rockfill dam (Aratozawa dam) suffered from a peak acceleration of about 10.24 m/s^2 recorded by strong motion accelerometers at its bottom gallery, which resulted in: (i) the significant increase of the pore water pressure in the mid-core; and (ii) a huge settlement at the crest of the dam (Ohmachi and Tahara 2011). The relevant numerical case study of the dam under two-component Miyagi earthquake loading reconfirmed that the brutal increase of excess pore water pressure was the most likely reason to explain the settlement observed at the crest (Zhu et al. 2020).

1.2 Overview of the Mechanical Behaviour of Sands

As a three-phase granular material, sand specimens can be generally composed of two elements: (i) solid grains forming the overall soil matrix and (ii) internal voids space involved in solid matrix which could be filled either by air or water or both. For fully saturated sand ($S_r = 100\%$), the internal voids space is totally filled by interstitial water and soil becomes then binary (sand matrix + pore water). In that case, these materials could be assimilated to a continuous medium for which a total stress tensor could be defined: σ_{ij}. According to Terzaghi's principle, the total stress tensor can be viewed as the sum of two distinct proportions: (i) an effective stress tensor σ'_{ij} considering the stress chain between sand grains; (ii) an isotropic tensor $u \bullet \delta_{ij}$ considering the stress sustained by interstitial water: $\sigma_{ij} = \sigma'_{ij} + u \bullet \delta_{ij}$, where δ_{ij} stands for the Kronecker symbol. The parameters employed for further analysis are summarized in Table 1.1.

1.2.1 Triaxial Loading and Drainage Conditions

For laboratory research, the most commonly used apparatus in soil mechanics is the triaxial setup, where a biaxial evolution can be expected since two of the three principal stresses are always identical, whether subjected to compression ($\sigma_2 = \sigma_3$ in Fig. 1.2a) or extension ($\sigma_1 = \sigma_2$ in Fig. 1.2b) loading.

Table 1.1 Fundamental variables for triaxial testing employed in the present work

Major principal stress	σ_1
Intermediate principal stress	σ_2
Minor principal stress	σ_3
Deviator stress	$q = \sigma_1 - \sigma_3$
Total mean stress	$p = (\sigma_1 + \sigma_2 + \sigma_3)/3$
Effective mean stress	$p' = (\sigma'_1 + \sigma'_2 + \sigma'_3)/3$
Major principal strain	ε_1
Intermediate principal strain	ε_2
Minor principal strain	ε_3
Deviator strain	$\varepsilon_d = 2(\varepsilon_1 - \varepsilon_3)/3$
Volumetric strain	$\varepsilon_v = (\varepsilon_1 + \varepsilon_2 + \varepsilon_3)/3$
Pore water pressure	u
Initial pore water pressure (back prcssurc)	u_0
Excess pore water pressure	$\Delta u = u - u_0$
Isotropic consolidation stress	σ'_c
Excess pore water pressure ratio	$r_u = \Delta u/\sigma'_c$

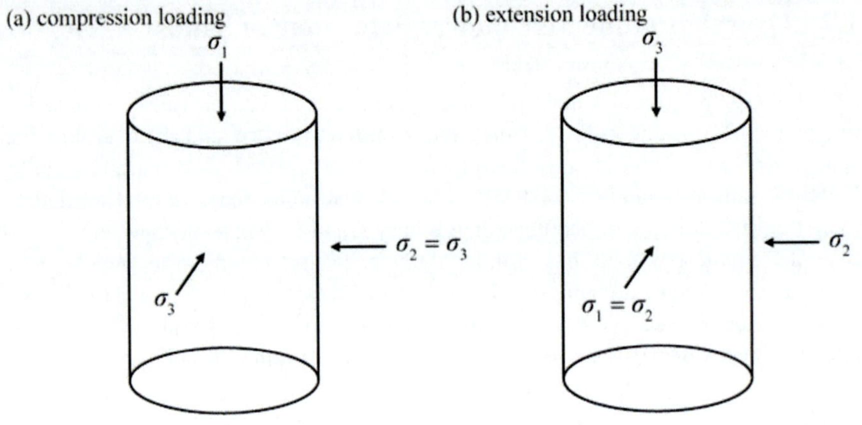

Fig. 1.2 Principal stress conditions upon triaxial testing: **a** compression loading; **b** extension loading

According to the drainage condition, triaxial tests for sandy specimens can be, in general, divided into two types during the shearing stage: (i) drained and (ii) undrained tests. In the first case, the drainage condition is set by opening the bottom drainage line; thereby, pore water is free to move in and out of the specimen through that line. In other words, upon shearing, there is no change in pore water pressure ($\Delta u = 0$) and the pore water pressure remains equal to its initial value u_0 (back pressure). Note that for these conditions, the drainage of pore water allows the occurrence of volumetric variation of the specimen (ΔV or $\varepsilon_v \neq 0$).

In the second case, both the top and bottom drainage lines are closed (see Fig. 1.3) during shearing. Hence, the flux of pore water is prevented (ΔV or $\varepsilon_v = 0$) resulting in the development of excess pore water pressure ($\Delta u \neq 0$). Thus, it is necessary to adopt the definition of effective stresses (defined in the above section) since their values can be much different from those of the total stresses.

1.2.2 Contractancy and Dilatancy of Sand Under Shear

As a granular material, the shear resistance of sand is mainly provided by between-grains occlusal frictions limiting the relative displacement or rolling of adjacent grains. In the case of loose sand (Fig. 1.4a), there is an irreversible tendency to a volume decrease of the specimen ($\varepsilon_v > 0$) upon shearing, which is referred to as "contractancy". In the case of dense sand (Fig. 1.4b), this trend turns into the opposite standing for a volume increase of the specimen ($\varepsilon_v < 0$), which is termed as "dilatancy".

An illustrative example concerning contractancy and dilatancy of sand was provided by Benahmed (2001). As for Hostun sand RF in a loose state with density

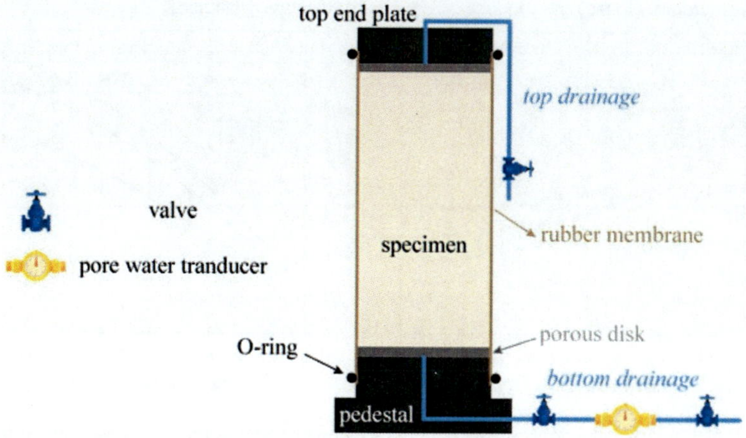

Fig. 1.3 Drainage conditions in a triaxial test

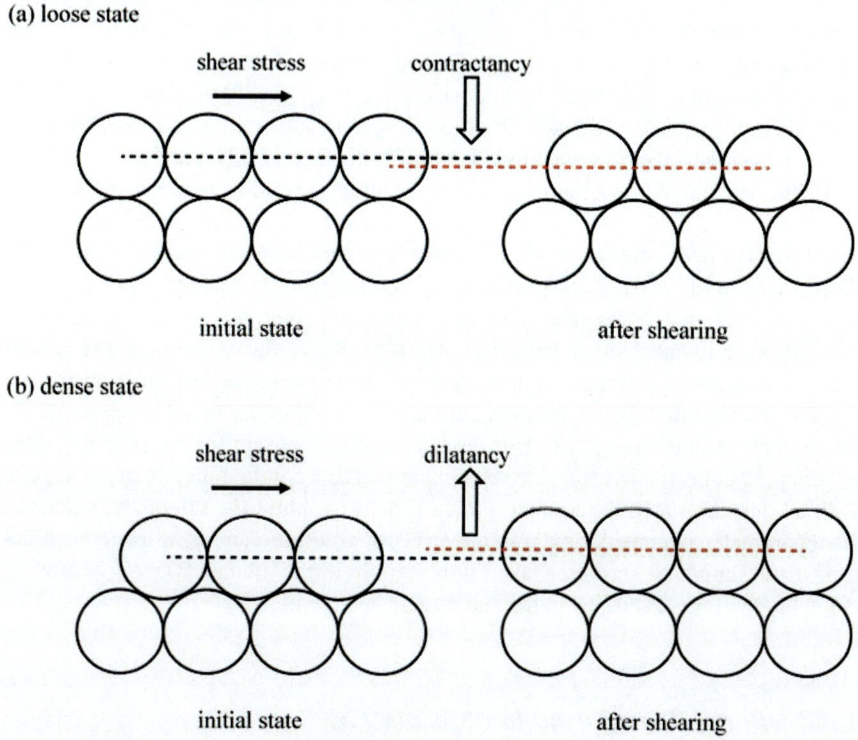

Fig. 1.4 Volume variation of a granular medium (schematic): **a** contractancy for an initially loose arrangement; **b** dilatancy for an initially dense arrangement

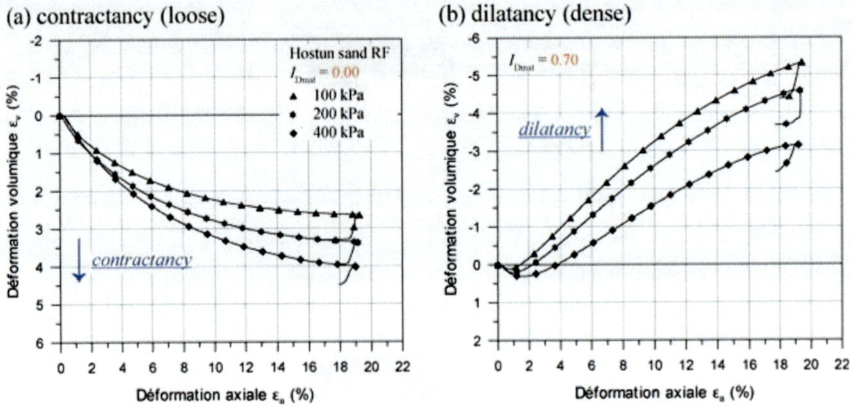

Fig. 1.5 Volumetric behaviours of Hostun sand RF upon triaxial testing: **a** a loose state of $I_{\text{Dmat}} = 0.00$; **b** a dense state of $I_{\text{Dmat}} = 0.70$ (after Benahmed 2001)

index $I_{\text{Dmat}} = 0.00$ under three values of the consolidation stress (see Fig. 1.5a), the test specimens continued to contract and finally achieved an asymptotic line. As for the same sand in a dense state with density index $I_{\text{Dmat}} = 0.70$ under the same three consolidation stresses (see Fig. 1.5b), the test specimens initially underwent a minor phase of contractancy, followed by a predominant stage of dilatancy.

In the triaxial stress–strain space, the contractancy and dilatancy of soil are commonly described as the incremental rate of the plastic volumetric strain to that of the plastic deviator strain: $d = \dot{\varepsilon}_v^p / \dot{\varepsilon}_d^p$, where "over dot" and superscript "p" stand for "increment" and "plastic", respectively. In the framework of advanced constitutive modelling, a better representation of sand dilatancy is then one of the critical issues in mimicking its mechanical behaviour. On the basis of the experimental results of clayey soil, with the assumption of incremental plastic energy in the Cam-Clay model, Roscoe and Schofield (1963) proposed dilatancy $d = M - q/p' = M - \eta = d(M, \eta)$, where η represents stress ratio. However, the above equation for cohesive soil does not adapt to cohesionless soil since d is expressed as a unique function of stress ratio with a constant material parameter M. In order to capture the dilatancy of sand, a state parameter ψ considering both the effect of void ratio and isotropic consolidation stress should be seriously taken into account in the dilatancy law $d = d(M, \eta, \psi)$, which will be discussed in the subsequent section of 1.3.1.

1.2.3 Phase Transformation State (PTS)

Based on the inversion of volumetric strain in which a soil element passes from contractancy to dilatancy, Ishihara et al. (1975) formulated the concept of phase transformation state (PTS) to translate the specific state where the rate of ε_v ($\dot{\varepsilon}_v$) or the rate of Δu ($\Delta \dot{u}$) is equal to zero in drained or undrained conditions, respectively. Two sets of monotonic triaxial tests showing the key characteristics of the phase transformation state were conducted by Benahmed (2001) under both drained and undrained conditions in which the red points in the graph could define the achievement of this state (Fig. 1.6).

For a given sand, this threshold can be supposed to be intrinsic since it is irrespective of the imposed loadings, density indexes and drainage conditions. Furthermore, its value is associated with a characteristic stress ratio (η_{car}) and a friction angle (ϕ'_{car}). In the q–p' space, the PTS line divides the stress space into three domains (Phong 1980): (i) contractive response if η^{+}_{car} (extension) < η < η^{+}_{car} (compression); (ii) dilatant response if $\eta > \eta^{+}_{car}$ (compression) and (iii) dilatant response if $\eta < \eta^{-}_{car}$ (extension, see Fig. 1.7).

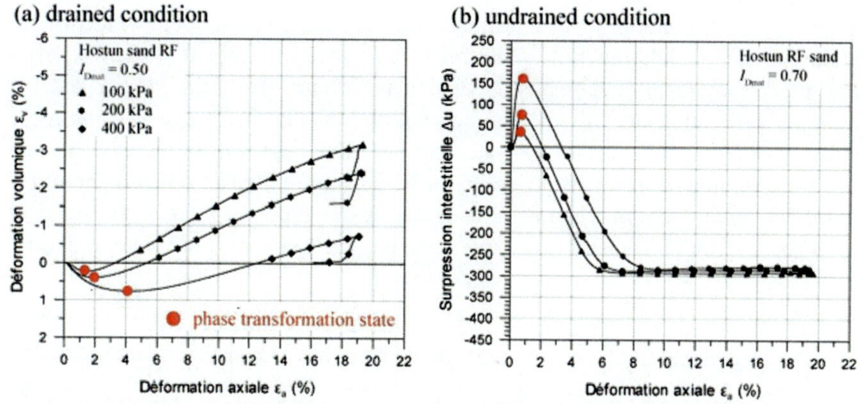

Fig. 1.6 Phase transformation states (red points) for Hostun sand RF: **a** case of drained condition; **b** case of undrained condition (after Benahmed 2001)

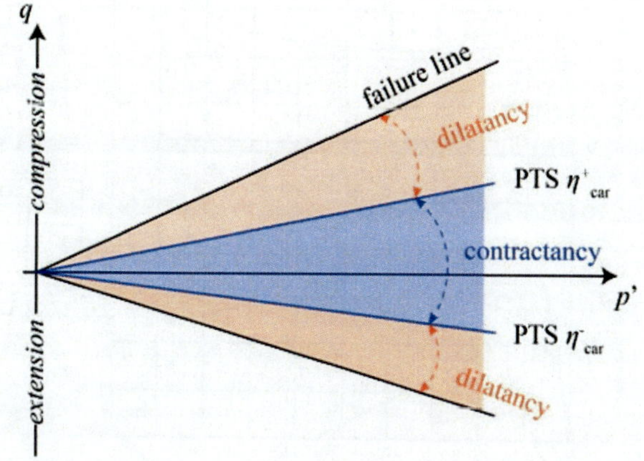

Fig. 1.7 Phase transformation state lines in the q–p' space (after Phong and Sidaner 1981)

1.3 Liquefaction of Pure Sand

1.3.1 Sand Liquefaction Under Monotonic Shear

In the literature, systematic laboratory experimental investigations concerning sand liquefaction can be traced back to the 1960s. Castro (1969) performed a series of undrained triaxial compression tests on Banding sand. Figure 1.8 shows the experimental results in which three distinct behaviours can be observed.

For test **a** in Fig. 1.8, a highly liquefiable response is justified by the following two factors: (i) the sudden decrease in deviator stress (q) was found after achieving a peak value; (ii) the excess pore water pressure (Δu) monotonously increased without any drop trend and reached the highest value very close to the initial consolidation stress σ'_c. For test **b** in Fig. 1.8, a limited liquefaction behaviour could be identified: (i) the specimen could regain its undrained shear resistance (about $\varepsilon_a = 10\%$) after the initial decease in deviator stress, which is different from the case **a** and (ii) the rapid generation of Δu was inhibited and the dilatant behaviour became more apparent making the excess pore water pressure rate negative. In the case of **c** in Fig. 1.8, a dilatant response is observed for which q presented an inflexion point for $\varepsilon_a \approx 2.5\%$ linking to a maximum value of Δu. Afterwards, the increase/decrease of $q/\Delta u$ became both increasingly conspicuous. Regarding excess pore water pressure,

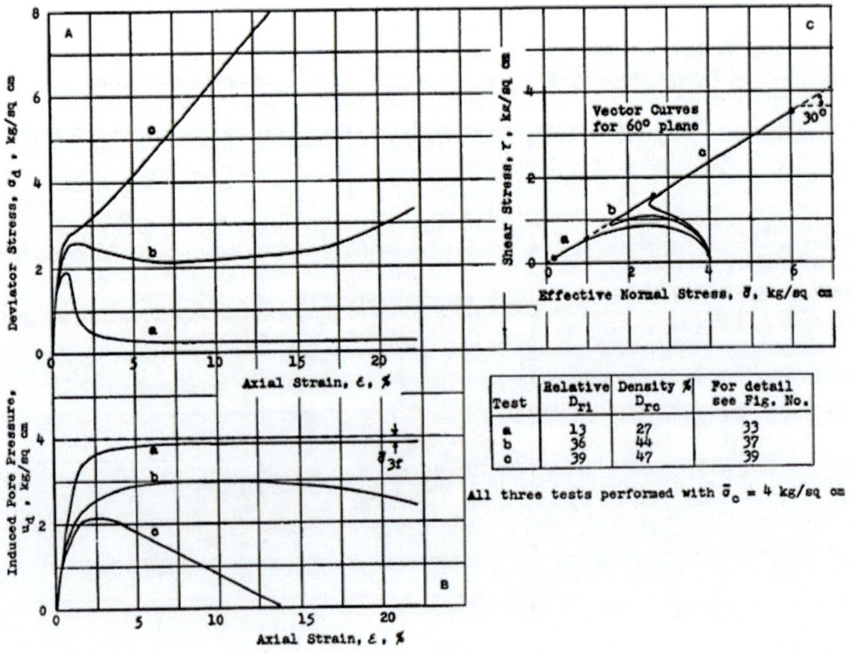

Fig. 1.8 Different behaviours for Banding sand under undrained conditions (from Castro 1969)

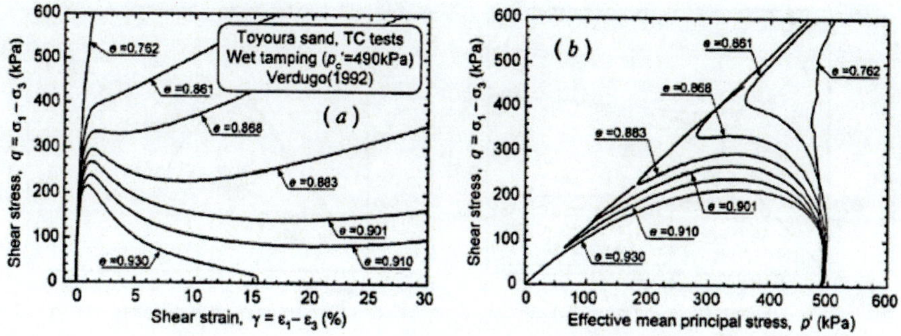

Fig. 1.9 Influence of density index on undrained behaviour of Toyoura sand: **a** stress–strain curves; **b** effective stress paths (from Yoshimine and Ishihara 1998)

it is worth noting that Δu could even become lower than atmospheric pressure at large ε_a.

1.3.1.1 Influence of Density Index

Yoshimine and Ishihara (1998) presented the results of undrained triaxial compression tests realized by Verdugo (1992) on Toyoura sand of distinct void ratios under a given consolidation stress of 490 kPa (Fig. 1.9). As for sand in loose state (e.g., void ratio $e = 0.930$ and 0.910), a very contractive response was observed, which can be justified by the sharp decrease in deviator stress (q). This is consistent with Castro (1969) as previously mentioned in the last paragraph. With the decrease in e, the contractive response was gradually substituted by dilatancy by passing through an intermediate limited liquefaction behaviour (e.g., void ratio $e = 0.883$). In a very dense state corresponding to $e = 0.762$, both p' and q steadily increased without any downward trend, exhibiting a strong strain-hardening dilatant behaviour. Regarding the peak value of deviator stress ($q_{\max}$), it should be noticed that $q_{\max}$ increased with the decrease of e, implying that liquefaction potential gradually disappeared. The above results are consistent with the experimental work performed by Kramer and Seed (1988); Canou (1989); Konrad et al. (1991) and Benahmed (2001).

1.3.1.2 Influence of Isotropic Consolidation Stress

For a given value of the void ratio e, the level of isotropic consolidation stress σ'_c was found to be another predominant factor influencing sand behaviour. In Fig. 1.10, Yoshimine and Ishihara (1998) presented the results of undrained triaxial compression tests realized by Verdugo (1992) on Toyoura sand specimens with the same void ratio under different consolidation stresses. Within the lower range of $\sigma_c' = 98$ and 981 kPa, it is found that (i) the stress–strain curves were not characterized by a clear peak point, as shown in Fig. 1.10a; (ii) the effective stress paths (ESP) in the q–p'

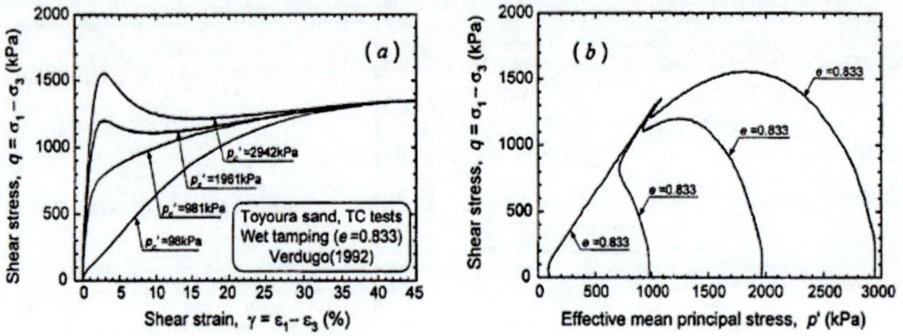

Fig. 1.10 Influence of consolidation stress on undrained behaviour of Toyoura sand: **a** stress–strain curves; **b** effective stress paths (from Yoshimine and Ishihara 1998)

space (Fig. 1.10b) indicate a dilatant behaviour since $p^{'}$ remarkably increased after an inflexion point. On the other hand, in the case of higher consolidation stress (e.g., $\sigma_c^{'} = 1961$ and 2942 kPa), the peak point of deviator stress apparently appeared and became, in particular, more important with the increase in $\sigma^{'}_{\mathrm{c}}$. Meanwhile, the ESPs described in Fig. 1.10b also suggest an increasingly enhanced contractive response. Both the above two phenomena reveal that the contractive behaviour of sand could be largely boosted by setting a higher level of $\sigma^{'}_{\mathrm{c}}$.

1.3.1.3 Concept of Critical State Line

While studying sand behaviour, the combined influence of e and $\sigma_c^{'}$ needs to be considered since even a minor variation of these two parameters can lead to a significant discrepancy in the corresponding mechanical response (see Figs. 1.9 and 1.10). In the pioneering work of Casagrande (1936), strain-controlled drained triaxial tests were performed on both initially loose and dense sand specimens, which formed the cornerstone of modern comprehension of soil mechanics. The results revealed that all specimens tested at the same value of $\sigma^{'}_{\mathrm{c}}$ approached approximately the same density when sheared to large strains: (i) initially loose sand (e_{L} in Fig. 1.11b) contracted upon shearing, or densified; (ii) initially dense sand (e_{D} in Fig. 1.11b) first contracted in a minor deformation range and then began to strongly dilate. In addition, at large strains, all specimens continued to be sheared with a constant deviator stress (Fig. 1.11a). The void ratio corresponding to this constant state is termed as "critical void ratio" e_{c}. By performing triaxial tests with varying values of $\sigma_c^{'}$, Casagrande further observed that e_{c} was only dependent upon $\sigma_c^{'}$.

In this sense, a critical state line (CSL) linking various initial states could be defined in the e-log$p^{'}$ space (see Fig. 1.12) so as to sketch the boundary between initially contractive and dilatant potentials. Since the CSL is able to delineate between contractive and dilatant behaviours, this line is afterwards used to capture the volume change trend in undrained conditions of saturated sand during shearing (see Fig. 1.12): (i) sand specimens with the initial state (e and $\sigma_c^{'}$ employed) plotting

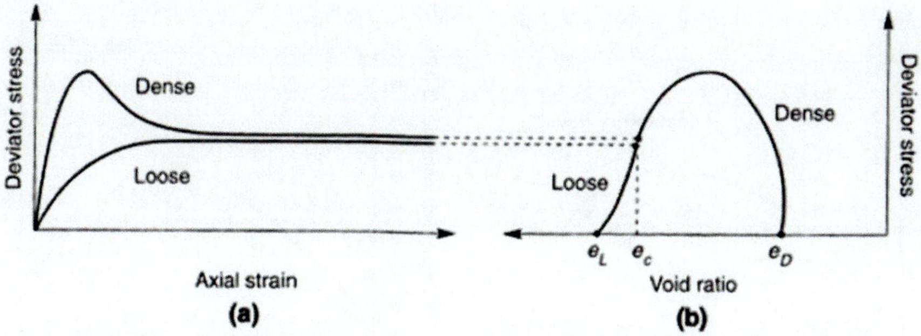

Fig. 1.11 **a** Stress–strain curves; **b** stress-void ratio curves for loose and dense sand under the same consolidation stress (from Casagrande 1936)

above the CSL is supposed to be highly susceptible to flow liquefaction; (ii) specimens of initial state plotting below the CSL is, on the contrary, insensitive to flow liquefaction. Based on the conception of CSL, a dimensionless scalar $\psi = e_{\text{cur}} - e_{\text{c}}$ was firstly introduced by Been and Jefferies (1985), which estimated the distance between the current and critical void ratios under the same value of σ'_{c}. Consequently, sand response could be considered as a function of ψ: (i) sand contracts while $\psi >$ 0 (in loose state); (ii) sand dilates while $\psi < 0$ (in dense state). Note that in order to avoid the computational problem of positive and negative signs, some researchers (Luo and Zhang 2004; Zhu and Cheng 2020) recently proposed using the ratio of the current and critical void ratio $\psi = e_{\text{cur}}/e_{\text{c}}$ instead of the difference measured in the *e*-log*p*' plane to outline the contractancy/dilatancy of sand in different states.

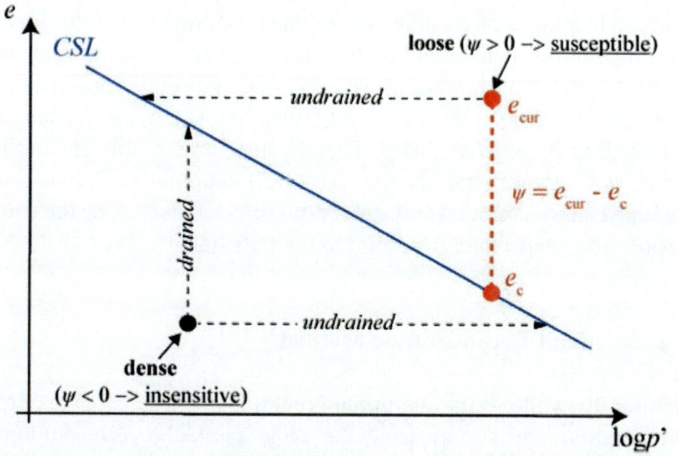

Fig. 1.12 Behaviour of initially loose and dense sand specimens under drained and undrained conditions for logarithmic consolidation stress scale (after Kramer 1996)

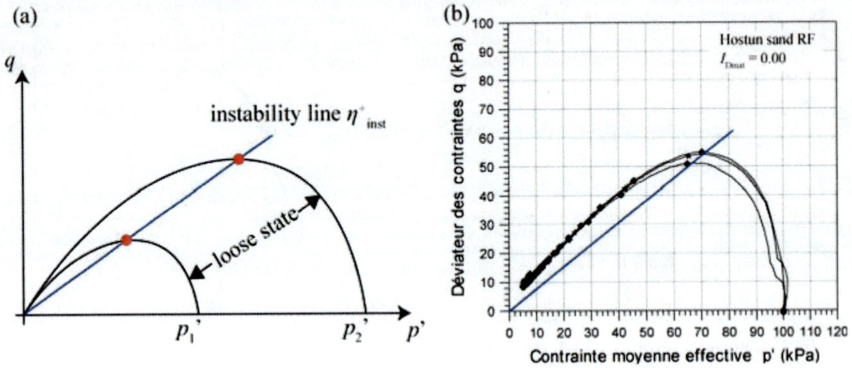

Fig. 1.13 Concept of instability line in the q-p' space (after Lade 1992): (a) conceptual scheme; (b) instability line for Hostun sand RF in a very loose state of $I_{\text{Dmat}} = 0.00$ (after Benahmed 2001)

1.3.1.4 Concept of Instability Line

As for a loose sand specimen with a highly contractive tendency subjected to undrained monotonic loading, the peak point of the effective stress path (red points in Fig. 1.13a) triggering flow liquefaction was experimentally found to pass through the origin of axes by a straight line. This line was first termed as the "instability line" by Lade (1992, 1993). An illustrative example concerning this concept was provided by Benahmed (2001) through examining Hostun sand RF in a very loose state of $I_{\text{Dmat}} = 0.00$, as shown in Fig. 1.13b.

Later, a detailed experimental study was performed by Benahmed (2001), proving that the instability line is not unique such as the PTS line who is intrinsic (irrespective of imposed loading, density indexes, drainage conditions) and this line normally rotates as a function of density index. In addition, it is widely recognized that the test specimens with a constant initial density index but under different consolidation stresses achieve almost the same instability line (Lade and Yamamuro 1997; Nguyen 2014; Benahmed et al. 2015). Thereby, the slope of this line has become a key indicator for evaluating the relevant liquefaction potential. The measure of the instability state can be identified by a stress ratio (η^{+}_{inst} and η^{-}_{inst} on the compression and extension side, respectively), similar to the PTS line.

1.3.1.5 Influence of Reconstitution Method

More importantly, while fixing the initial conditions of triaxial tests (e.g., density index and consolidation stress) using the same sand, the reconstitution method itself has also a significant impact on sand response (Mulilis et al. 1977): among dry tamping (DT) method (Benahmed et al. 2004), moist tamping (MT) method (Casagrande 1976; Ladd 1978; Frost and Park 2003), dry pluviation (DP) method

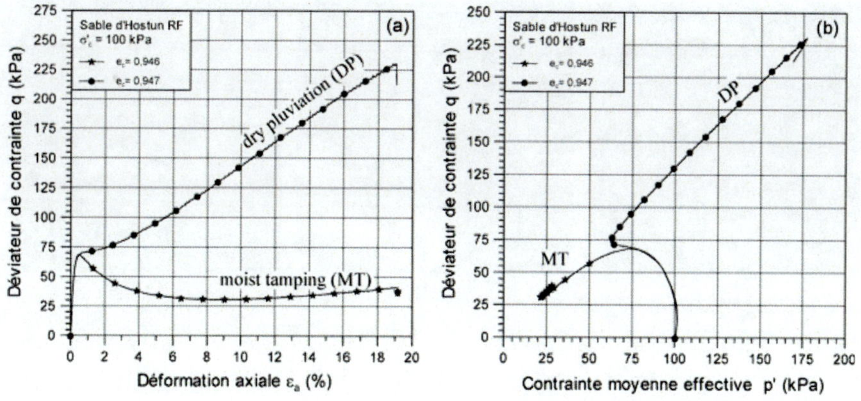

Fig. 1.14 Influence of the reconstitution method (MT vs DP) on the behaviour of Hostun sand RF in a loose state: **a** stress–strain curves; **b** effective stress paths (after Benahmed et al. 2004)

(Kolbuszewski 1948; Levacher and Garnier 1994; Benahmed et al. 2004), etc.[1] In other words, the manner in which a sand specimen is prepared to a given density index has a significant influence on the initial soil fabric, controlling its subsequent response. Figure 1.14 presents the results of undrained triaxial tests on loose Hostun sand RF ($e_c \approx 0.946$, Benahmed et al. 2004), it can be seen that (i) while applying the MT method, the sand response is very contractive, directly characterized by both the significant decreases of $q/p^{'}$ in stress–strain curve and effective stress path; (ii) whilst a dilatant response is observed for the specimen with approximately the same void ratio ($e_c = 0.947$ vs 0.946) prepared by the DP method.

The above significant discrepancies may be interpreted in terms of the initial soil fabric linking to different soil reconstitution methods. The moist conditions created by the MT method, which consists of adding 5% of water to total dry mixture by weight, create a wet condition where sand grains are generally aggregated forming a so-called "metastable" structure with the presence of macropores, as shown in Fig. 1.15a, b. These macropores, upon shearing, may collapse, which results in a very contractive response and a high generation of Δu when sheared in undrained conditions. Note that this "metastable" structure was previously referred to as "honeycomb-structure" or "bulked sand" by Casagrande (1976). On the other hand, in dry conditions created by the DP method where sand grains are much uniformly arranged, the overall sand matrix against external loadings is regularly placed without the presence of macropore (see Fig. 1.15c), which certainly promotes a more strain-hardening dilatant response, as proved by Fig. 1.14.

[1] An essential summary of different sandy specimen preparation methods is given in Zhu et al. (2021a, b).

Fig. 1.15 Microphotographs of Hostun sand RF showing two distinct initial fabrics: **a–b** metastable aggregates and macropores (MT); **c** regular stack (DP) (from Benahmed et al. 2004)

1.3.2 Sand Liquefaction Under Cyclic Shear

Regarding engineering earthquake simulation in laboratory research, cyclic triaxial apparatuses certainly play a prevailing role (Tsukamoto and Ishihara 2022) since they are capable of producing a sequence of deviator stresses, very similar to seismic loading. For this purpose, the test specimens should be isotropically consolidated with consolidation stress ratio $K_c = 1$ ($K_c = \sigma'_a/\sigma'_r$ where σ'_a and σ'_r stand for axial and radial effective stress, respectively), ensuring that the test specimen can undergo an alternated equal-amplitude loading on its 45° plane (see Fig. 1.16) similar to the shear produced on the horizontal plane in the level ground during earthquakes for a saturated soil element. Under cyclic triaxial conditions, it is usual to take either the excess pore water pressure ratio r_u or axial strain ε_a as a yardstick to recognize the outset of liquefaction. Between these two criteria, Ishihara (1995) emphasized the priority of utilizing the axial strain ε_a since the excess pore water pressure Δu of sand containing fine particles (e.g., silty sand and clayey sand) was experimentally found not to totally develop till the initial confining pressure but rather stop building up when it reached a value equal to approximately 90–95% of the initial consolidation stress.

1.3.2.1 Cyclic Mobility

Sand liquefaction behaviour as studied in cyclic triaxial test corresponds to two main phenomena according to the density index of sand matrix I_{Dmat}: (i) cyclic mobility in both medium-dense and dense states and (ii) true liquefaction (or flow liquefaction) in very loose state. An illustrative example concerning cyclic mobility (Hostun sand RF of $I_{Dmat} = 0.50$) was given by Benahmed (2001), as shown in Fig. 1.17. As the cyclic axial stress is applied, it can be seen that the excess pore water pressure increases and gradually reaches an ultimate value equal to the initial consolidation stress of 200 kPa. During the initial loading stage in Fig. 1.17b, the axial strain remains very

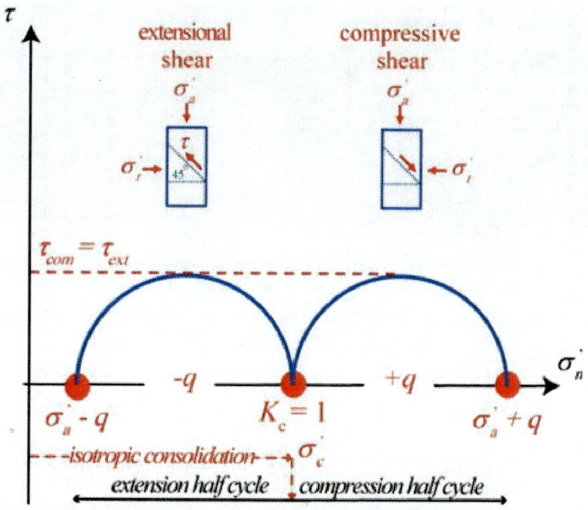

Fig. 1.16 Schematic diagram of equal-amplitude alternated loading on the 45° plane of test specimen in cyclic triaxial test (after Ishihara 1995)

small. Approaching the failure stage, further loading sharply deforms the specimen, yielding a fast-growing ε_a. Local zooms are provided in Fig. 1.17. After attaining a critical N_{cyc} of about 40 in the graph, the "double peak mechanism" with respect to excess pore water pressure curve becomes predominant. In the subsequent cycles, the excess pore water pressure *gradually* approaches the consolidation stress. At the same time, the axial strain significantly increases in Fig. 1.17d and the stress–strain curve takes a "reversed-z" shape in Fig. 1.17e, indicating that a sizable amount of energy dissipation occurs during a single cycle of loading. In Fig. 1.17f, the effective stress path continues to move left and finally passes through the original point twice in a cycle, exhibiting the so-called "butterfly wing" shape.

1.3.2.2 True Liquefaction (Or Flow Liquefaction)

Figure 1.18 provides an illustrative example of flow liquefaction for a very loose state (Hostun sand RF of $I_{Dmat} = 0.00$). As the cyclic axial stress is applied, the excess pore water pressure increases and finally attains a level equal to the consolidation stress of 200 kPa. In the initial loading stage in Fig. 1.18b, there is a minimal accumulation of axial strain, followed by a sudden drop. After triggering the flow liquefaction (e.g., N_{cyc} of about 114 in the graph), the excess pore water pressure *abruptly* increases to the consolidation stress rather than build up in a gradual manner, which is in contrast to that observed for cyclic mobility. At the same time, a complete collapse of the test specimen can be ascribed to the transient axial strain in the same cycle (see Fig. 1.18d). Owing to the outset of true liquefaction, the effective stress path in Fig. 1.18e goes immediately to the original point and the test specimen develops a liquid-like behaviour indicating a complete loss of shear strength.

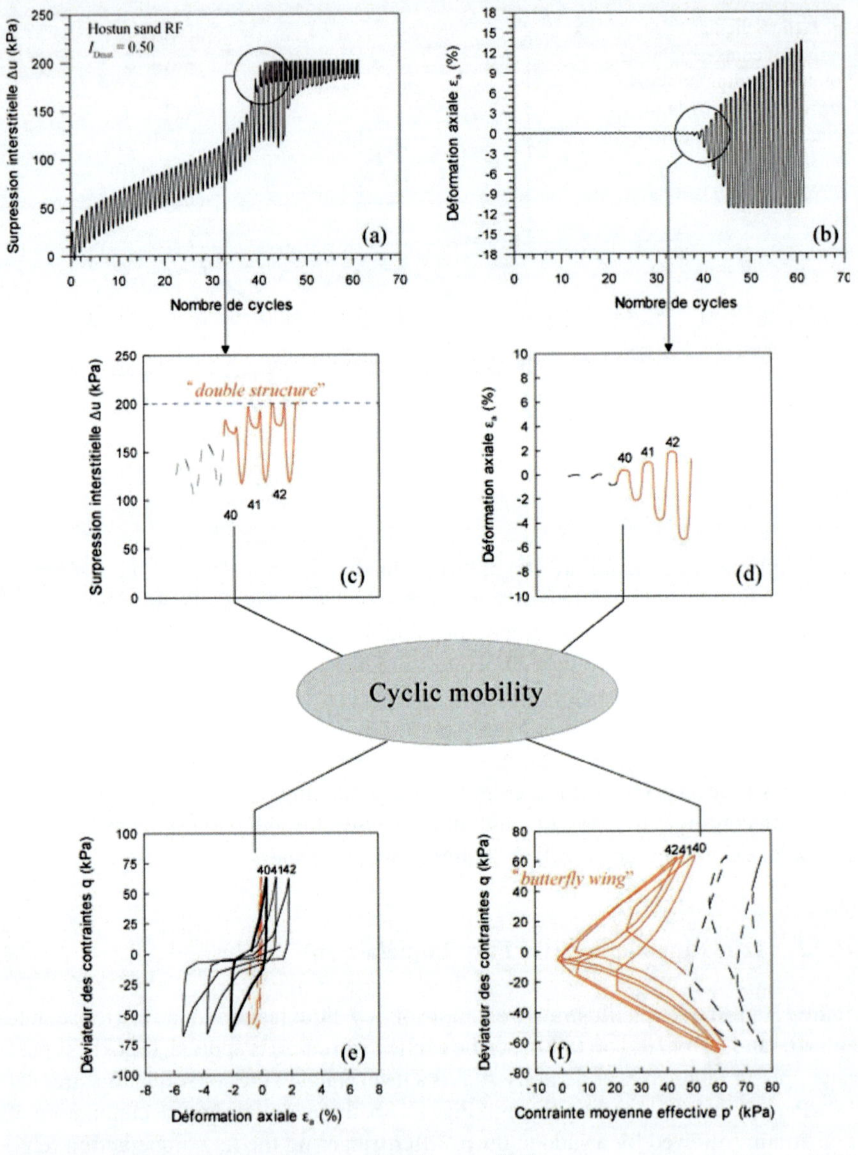

Fig. 1.17 Cyclic mobility for Hostun sand RF ($I_{\mathrm{Dmat}} = 0.50$): **a** excess pore water pressure curve; **b** axial strain curve; **c–f** local zooms (after Benahmed 2001)

1.3.3 Synthesis of Behaviours Observed Under Monotonic and Cyclic Shear

On the basis of a complete experimental database on Hostun sand RF, a synthetic work was finally derived by Benahmed (2001). The aim was to propose a single conceptual

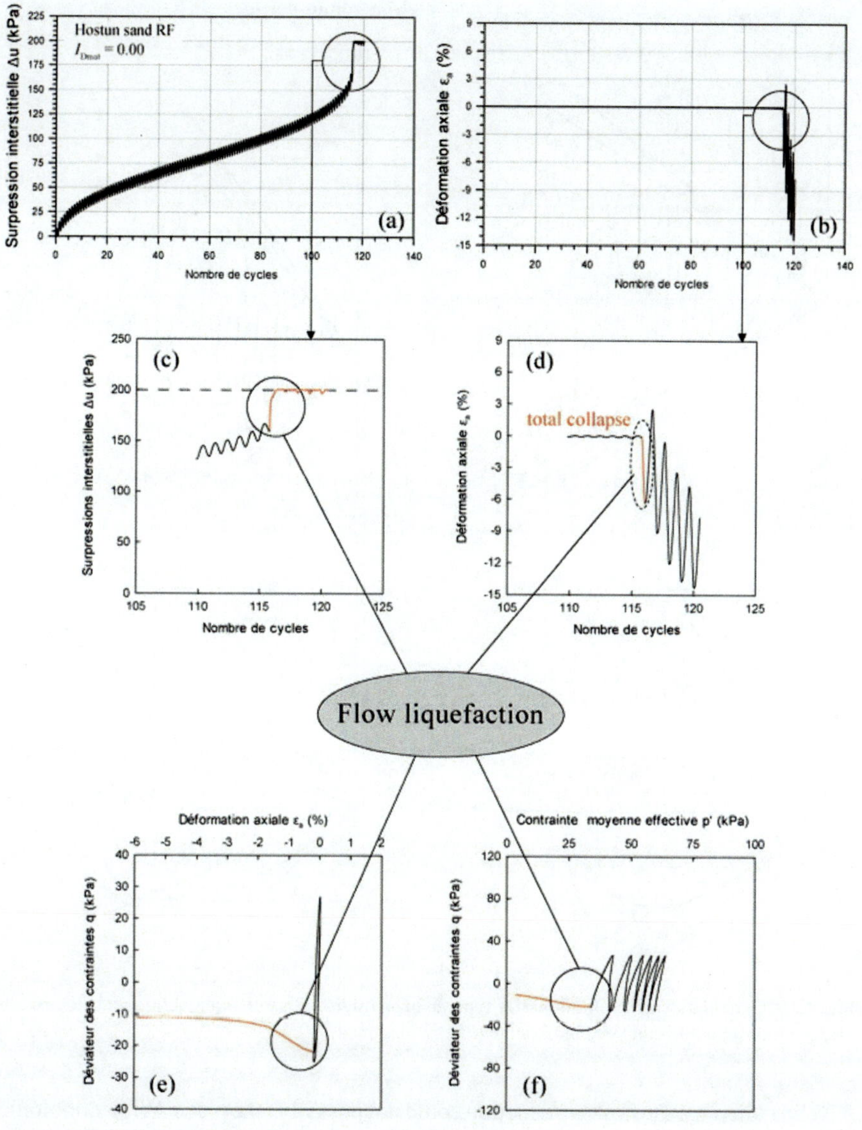

Fig. 1.18 True liquefaction for Hostun sand RF ($I_{\text{Dmat}} = 0.00$): **a** excess pore water pressure curve; **b** axial strain curve; **c**–**f** local zooms (after Benahmed 2001)

framework to synthesize all scenarios of sand liquefaction for both monotonic and cyclic loadings.

(i) **Very dense state in Fig. 1.19a**

For very dense state, sand specimen exhibits strong dilatant behaviour subjected to both compression and extension shearing. Its response is mainly determined by the

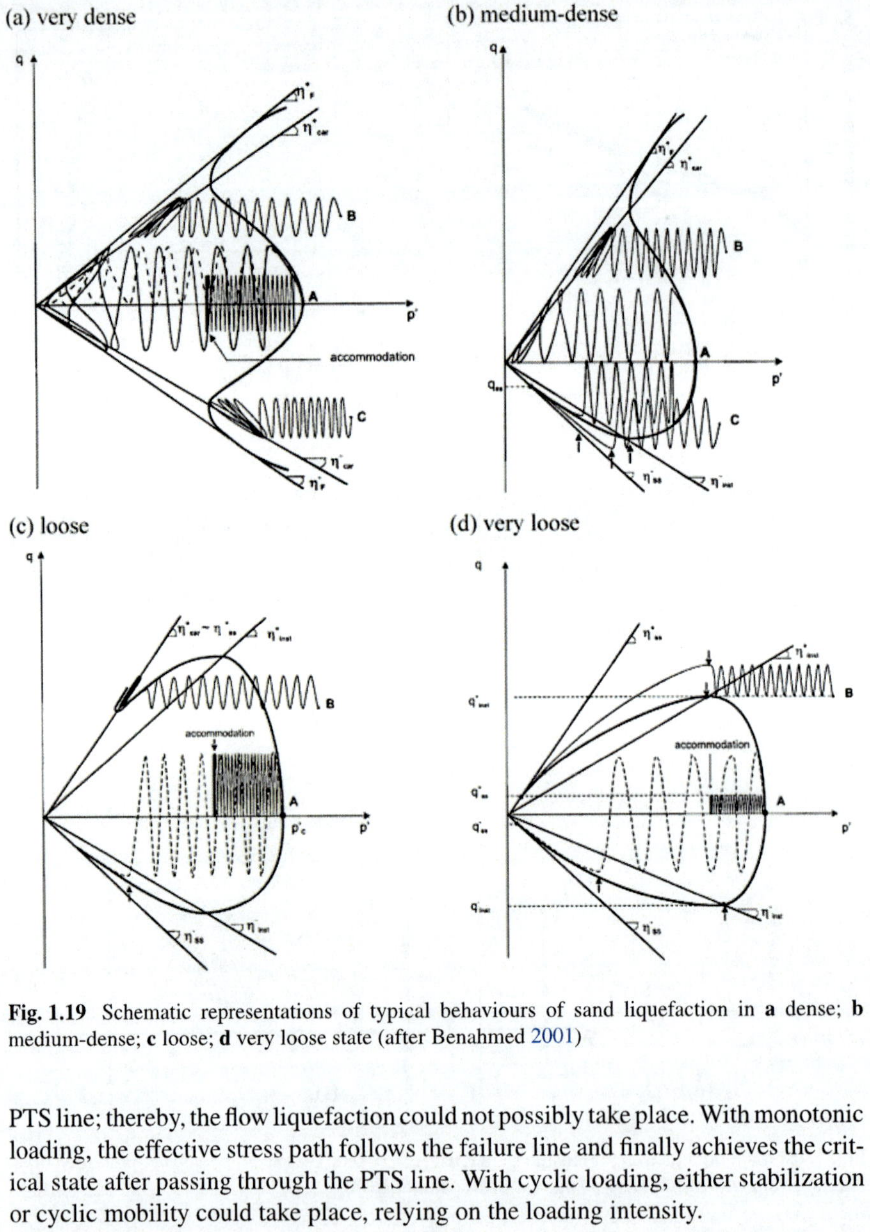

Fig. 1.19 Schematic representations of typical behaviours of sand liquefaction in **a** dense; **b** medium-dense; **c** loose; **d** very loose state (after Benahmed 2001)

PTS line; thereby, the flow liquefaction could not possibly take place. With monotonic loading, the effective stress path follows the failure line and finally achieves the critical state after passing through the PTS line. With cyclic loading, either stabilization or cyclic mobility could take place, relying on the loading intensity.

(ii) **Medium-dense state in Fig. 1.19b**

For medium dense state, sand specimen exhibits stable dilatant behaviour on the compression side (governed by the PTS line), whilst contractive response (liquefaction potential) is manifested on the extension side (governed by the instability line). In the first case, the corresponding effective stress path follows the same trajectory described in the very dense state. In the second case, the liquefaction can be triggered while the effective stress path passing through the instability line under either monotonic or cyclic loading.

(iii) **Loose state in Fig. 1.19c**

For loose state, sand specimen exhibits a high liquefaction potential (contractive behaviour) as being governed by the instability line subjected to triaxial compression or extension loading. With monotonic loading on the compression side, limited or intermediate liquefaction behaviour can be found (residual strength $\neq 0$), whilst total liquefaction collapse can be identified on the extension side.

(iv) **Very loose state in Fig. 1.19d**

For very loose state, sand specimen exhibits a boosted liquefaction potential and its response is entirely governed by the instability line. With monotonic loading on both sides, a complete static liquefaction response can take place with almost zero residual strength near the failure stage, in contrast to that observed in loose state. With cyclic loading, the true liquefaction can be triggered on the extension side.

1.4 Liquefaction Response of Sand Containing Fines

Over the last sixty years, sand liquefaction has become a very important topic of soil mechanics in relation to earthquake and anti-seismic engineering. Most of the earlier research focused on the liquefaction of clean uniform sand without considering the possible influence of fine particles. As a consequence, the understanding of clean sand liquefaction has already been acquired. However, loose liquefied sand in its natural state seems to so often contain more or less fine particles, which is frequently mentioned by various field investigations after the occurrence of sand liquefaction. For instance, after studying two destructive earthquakes that occurred in China, the Haicheng earthquake of 1975 and the Tangshan earthquake of 1976, Shengcong and Tatsuoka (1984) revealed that the presence of silt was very common at liquefied sites. Surface evidence of liquefaction was also observed for silty sand without previous treatment during the 1978 Miyagi earthquake in Japan (Ishihara et al. 1980). Regarding the quantity of fines, Ohsaki (1970) reported that soils with a fine fraction lower than 10% were more likely to liquefy during the 1964 Niigata earthquake. A similar conclusion was drawn by Tokimatsu and Yoshimi (1983) that more than half of all liquefied data fell within a small range of fines fraction (F_c) lower than 5% (Fig. 1.20) after analysing 17 worldwide earthquakes. Therefore, it is of great importance to investigate the undrained shear strength of loose sand with small-size inclusions (e.g., the addition of silt or clay) at relatively low fines fractions F_c (< 10%) while facing liquefaction disasters.

Over the past three decades, quite a few experimental investigations related to the effect of small-size particles (silt or clay) on the undrained shear strength of sand

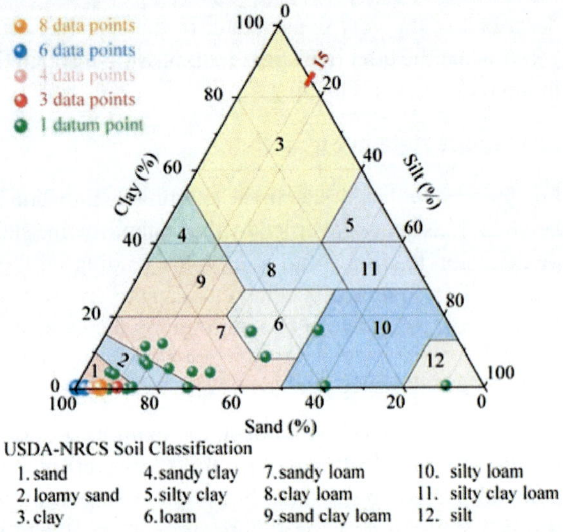

Fig. 1.20 Fines fraction of liquefied soils in a study of 17 worldwide earthquakes (after Tokimatsu and Yoshimi 1983)

mixtures have been reported. However, no clear consensus has been achieved and diverse opinions still exist as to whether the effect of fines is adverse or beneficial to undrained shear strength. In the literature published, with increasing fines fraction, the undrained shear strength of sand-fines mixtures could either: (i) increase (Seed et al., 1983, 1985; Pitman et al. 1994; Amini and Qi 2000; Ni et al. 2004; Benahmed et al. 2015; Jradi 2018); (ii) decrease (Erten and Maher 1995; Polito 1999; Bouferra and Shahrour 2004; Ghahremani et al. 2006; Belkhatir et al. 2010; Stamatopoulos 2010) or (iii) initially decrease until a certain threshold and then increase (Lade and Yamamuro 1997; Thevanayagam et al. 2002; Xenaki and Athanasopoulos 2003; Yang et al. 2006; Abedi and Yasrobi 2010). Efforts have been made to explain the above at least controversial or even contradictory results in the following sections (from §1.4.1-§1.4.4).

1.4.1 *Influence of Plasticity Index*

Recently, the impact of plasticity index of fine particles on sand liquefaction behaviour has been carefully examined. In an experimental study reported by Jradi (2018) and Jradi et al. (2021), two fine particles with distinct plasticity indexes (a non-plastic silt C500 and a plastic clay Speswhite with plasticity index $I_p = 30$) were selected and mixed with a French reference sand (Fontainebleau sand) of $I_{Dmat} = 0.10$, respectively. For non-plastic C500 fines, the test specimens were reconstituted by the moist tamping method, whilst the dry tamping method was chosen for plastic Speswhite to ensure a proper homogeneity of the sand-fines mixtures. A series of monotonic and cyclic triaxial tests was respectively conducted. As for cyclic loading, the corresponding cyclic shear resistance curves are summarized in Fig. 1.21a where two phenomena should be noted: (i) the addition of non-plastic C500 fines provided a very positive contribution to the liquefaction resistance; (ii) by contrast, the addition of plastic Speswhite lowered the liquefaction resistance. As for monotonic loading, the stress–strain curves for test specimens containing two different types of fine particles are shown in Fig. 1.21b, allowing the same conclusion to be drawn.

Another valuable experimental work regarding the role of plasticity index of fines was presented by Pitman et al. (1994). With the addition of fine particles of distinct plasticity indexes (plastic kaolinite fines and non-plastic crushed quartz) using the moist tamping technique, the experimental results showed that the undrained shear resistance of Ottawa C-109 reference sand was enhanced in both the cases, as displayed in Fig. 1.22. After normalizing the deviator stress σ_{d} by the peak undrained shear strength (σ_{dp}), there appeared to be only a small difference using either plastic or non-plastic fines.

(a) cyclic loading

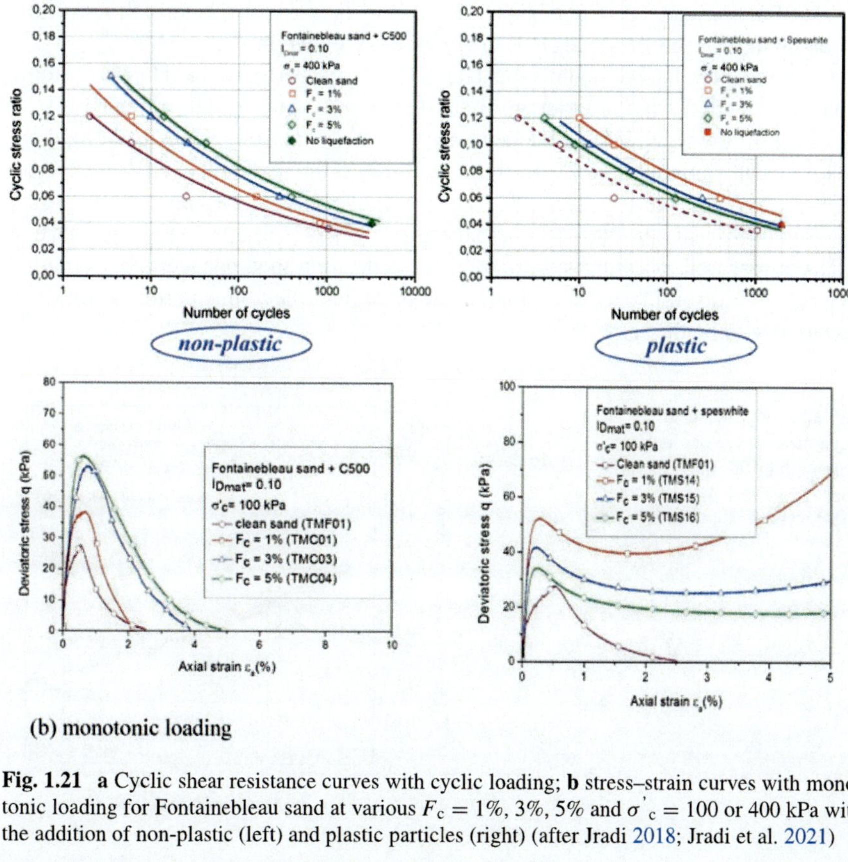

(b) monotonic loading

Fig. 1.21 **a** Cyclic shear resistance curves with cyclic loading; **b** stress–strain curves with monotonic loading for Fontainebleau sand at various $F_{\mathrm{c}} = 1\%$, 3%, 5% and $\sigma'_{\mathrm{c}} = 100$ or 400 kPa with the addition of non-plastic (left) and plastic particles (right) (after Jradi 2018; Jradi et al. 2021)

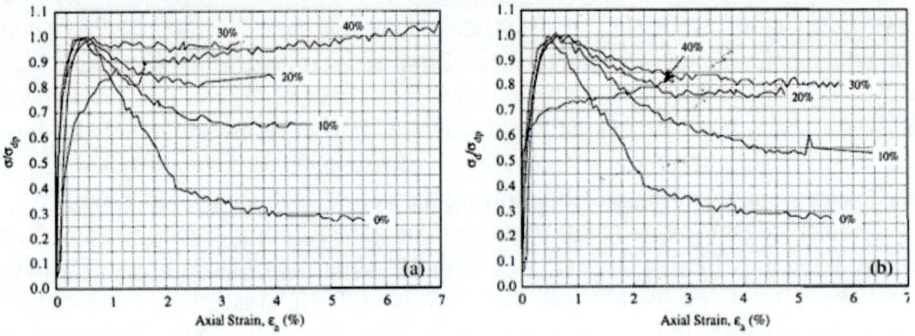

Fig. 1.22 Normalized deviator stress- strain curves on Ottawa C-109 reference sand with the addition of: **a** plastic kaolinite fines; **b** non-plastic crushed quartz (from Pitman et al. 1994)

The above finding is very similar with Park and Kim (2013) who pointed out that, in a very loose state with a high liquefaction potential, although the liquefaction resistance of sandy specimens indeed decreased with the increase in plasticity index of added materials (e.g., from $I_p = 8$ of a silt to $I_p = 50$ of a bento-silt used), the observed descending tendency was only slight, as shown in Fig. 1.23.

More recently, Papadopoulou and Tika (2016) have performed a set of monotonic triaxial tests (moist tamping method) and the experimental results revealed that in the range of low values of I_p of added fine particles (e.g., from NP to $I_p = 12$), there was no clear unique relation between I_p and q as a threshold plasticity index (I_{pth}) certainly existed (see Fig. 1.24). In fact, the above diverse viewpoints concerning the effect of plasticity index of fine particles might be perhaps due to the fact that different researchers employed different reconstitution methods, certainly creating different initial soil fabrics, for which the mechanical role of fine particles could not necessarily to be the same.

Fig. 1.23 Cyclic shear resistance curves for loose state with different types of fines (from Park and Kim 2013)

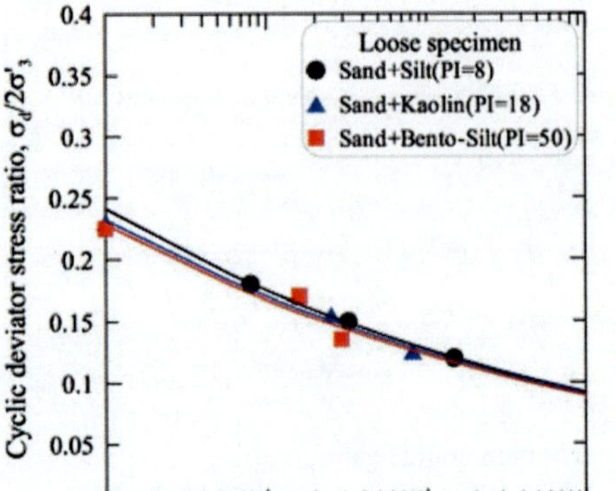

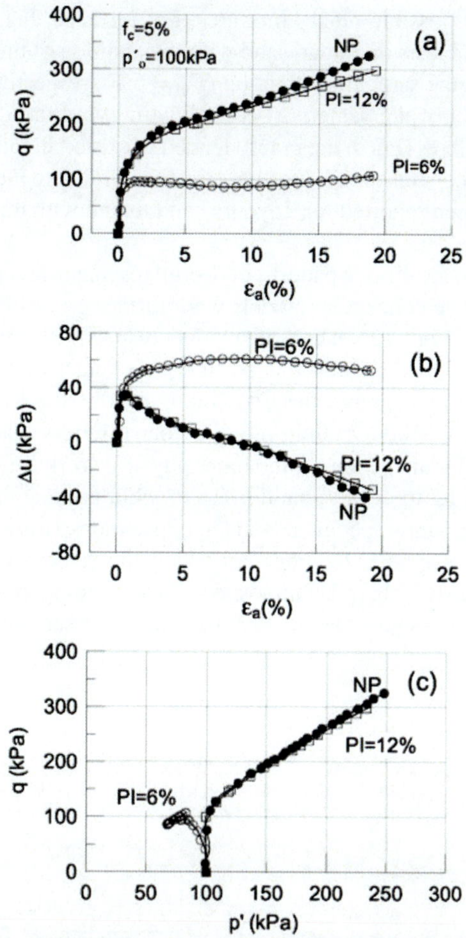

Test	PI (%)	Symbol	e	e_g	ε_{UI} (%)	ε_{QSS} (%)	ε_{PT} (%)	% clay/% silt
SF5 (NP)-3	NP	●	0.710	0.800	-	-	0.21	0.6/4.4
SF5 (PI=6)-3	6	○	0.709	0.799	1.25	7.91	9.27	0.9/4.1
SF5 (PI=12)-2	12	□	0.718	0.808	-	-	0.14	1.5/3.5

Fig. 1.24 Undrained monotonic triaxial compression tests for the sand-fines mixtures with F_c = 5% and σ'_c = 100 kPa: **a** stress–strain curves; **b** excess pore water pressure-axial strain curves; **c** effective stress paths (from Papadopoulou and Tika 2016)

1.4.2 Influence of Particle Shape

In order to clarify the influence of the shape of added particles, the experimental study performed by Yang and Wei (2012) involved two types of fines with distinct particle shapes (angular crushed silica of R = 0.377 and a rounded glass bead of R = 1, where R is a shape characteristic that represents the average angularity of the corners and edges of a grain) and two reference sands (Toyoura and Fujian sands), thereby, giving four possible binary mixtures. For instance, Fig. 1.25 schematizes the undrained triaxial tests on Fujian sand with the addition of the above-mentioned two fine particles at the same fines fraction ($F_c = 5\%$), respectively. It can be seen that, in the condition of the same post-consolidation void ratio, the test specimen containing rounded fines (FG in the graph) tended to exhibit higher flow liquefaction potential than the specimen containing angular fines. Similar to Fujian sand, the same tendency has also been reported for Toyoura sand mixed with the identical two fine particles.

In order to explain the diverse patterns of overall responses associated with particle shape, four grain-scale conceptual models were further proposed by Yang and Wei (2012), as shown in Fig. 1.26. The four models respectively represent the contact and movement of rounded/angular fine particles in contact with a rounded/angular coarse sand grain. In the graph, the two equal sides of the triangle simulate the path of a fine particle sliding/rolling along a coarse sand surface. In general, the more angular a sand grain becomes, the more physical work is required for the fine particle to move along the coarse sand grain to yield particle arrangement, which makes the composite sandy specimen (sand + plastic/non-plastic fines) significantly more resistant to the applied loading. Following the same logic, one might also expect that the shear strength of a composite soil can become much more significant if the added fine particles become more angular. In fact, a semblable conclusion was

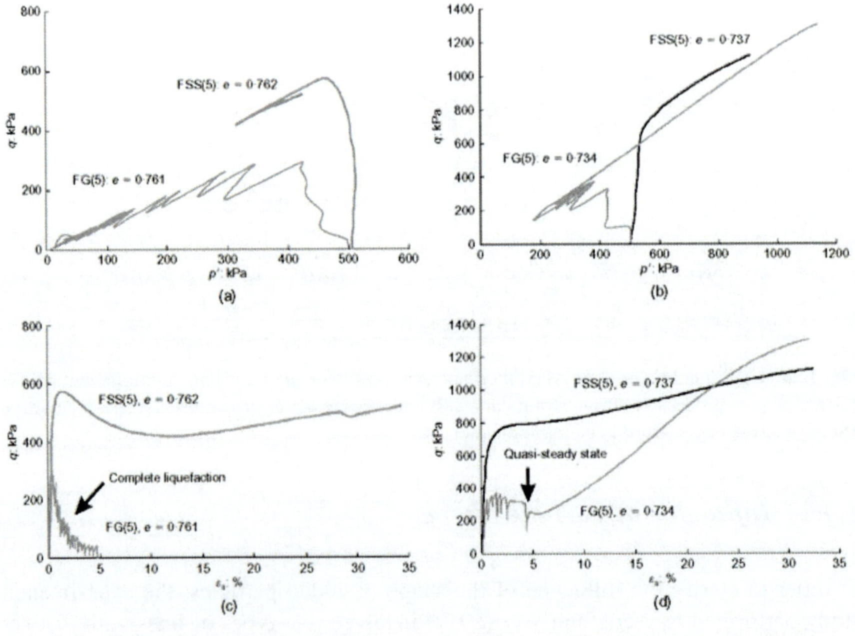

Fig. 1.25 Undrained shear behaviour of Fujian sand with the addition of fines of different shapes (FSS = Fujian sand mixed with angular crushed silica fines; FG = Fujian sand mixed with rounded glass beads): **a–b** effective stress paths; **c–d** stress–strain curves (from Yang and Wei 2012)

also drawn by Carraro et al. (2009) who reported an increase/decrease of both peak and critical-state friction angles by performing triaxial tests on a host sand with the addition of non-plastic silt/plastic kaolinite clay and pointed out that more angular non-plastic silt tended to increase the sliding resistance (enhancing effect) during shearing, whilst more rounded clay particles displayed a wakening effect.

However, a totally opposite configuration (see Fig. 1.27a) has been recently proposed by Monkul et al. (2017). At low fines fractions, the fine particles could locate between contact points of adjacent sand grains, certainly weakening the sand-to-sand contacts by separating them. This kind of contact between sand and fines is called "metastable contact", whereas the sand-to-sand contact yielding the global shear strength is called "stable contact". When subjected to shearing, fine particles

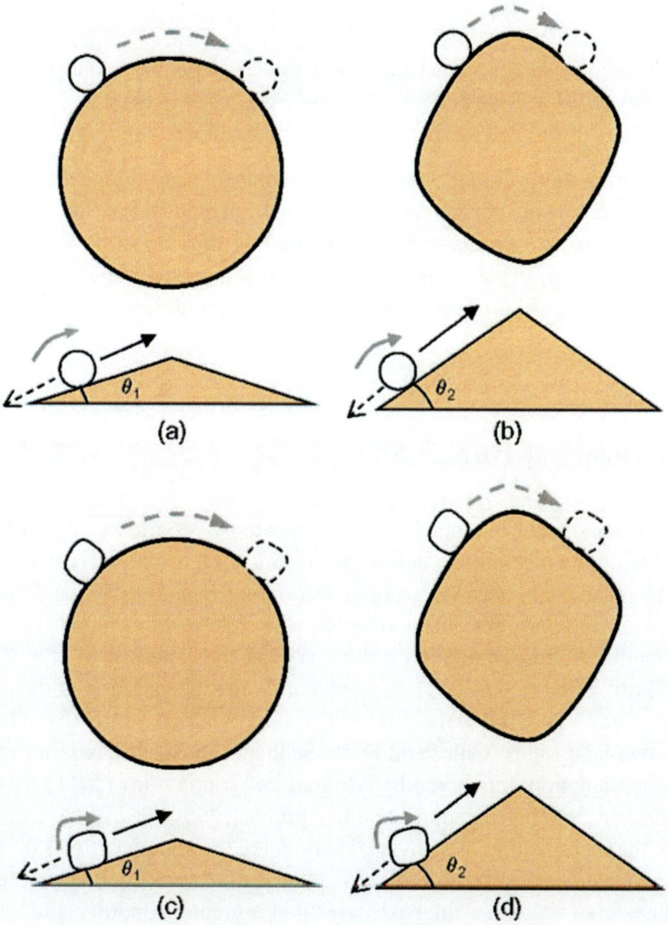

Fig. 1.26 Conceptual models explaining particle shape effects in mixed sand: **a** round-to-round; **b** round-to-angular; **c** angular-to-round; **d** angular-to-angular (from Yang and Wei 2012)

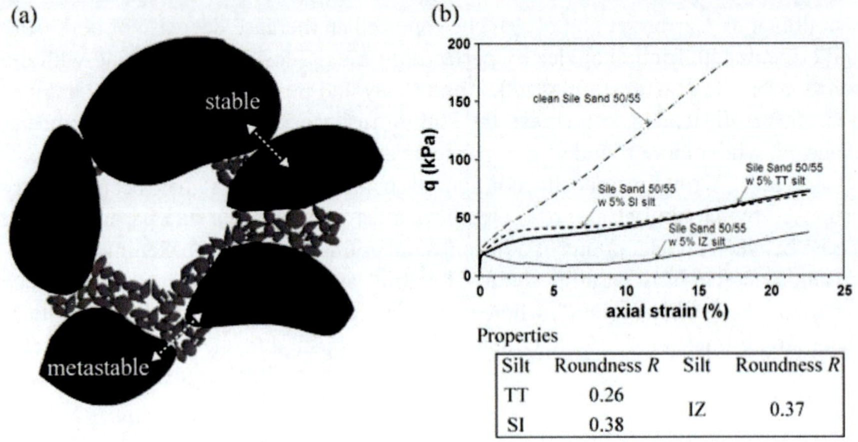

Silt	Roundness R	Silt	Roundness R
TT	0.26	IZ	0.37
SI	0.38		

Fig. 1.27 **a** Conceptual model of sand matrix with the addition of fine particles; **b** influence of roundness of fine particles on the stress–strain curves (after Monkul et al. 2017)

tend to move into intergranular voids between adjacent sand grains. Meanwhile, the global sand matrix would shrink with the sliding of fine particles. Such a mechanism behind contributes to a contractancy phenomenon, thus boosting the liquefaction potential. In addition, it was experimentally found that the angular nature of fine particles (see Fig. 1.27b) produced more metastable contacts, leading to an enhanced contractiveness.

1.4.3 Influence of Particle Size

Recently, the mean size of fines has been repeatedly reported to be an influential factor governing the mechanical behaviour of sand-fines mixtures (Liu et al., 2014). With the decrease of the various grain sizes (among D_{10}, D_{30}, D_{50} and D_{60}) due to the addition of fine particles, shear strength was observed to gradually decrease in the experimental finding of Belkhatir et al. (2011) and Janalizadeh et al. (2013). In addition, Cherif Taiba et al. (2016) succeeded in fitting the variation of the peak shear strength (q_{peak}) as a function of the different grain sizes so as to unify the experimental results for two host sands: Chlef and Fontainebleau sands, as shown in Fig. 1.28. A reasonable explanation was given by Monkul and Yamamuro (2011) by using the mean diameter ratio between coarse and fine grains ($D_{50\text{-sand}}/d_{50\text{-silt}}$) as a comparison basis. With the increment in this physical parameter, the added particles became finer and mostly located in the intergranular voids instead of forming metastable contacts (sand-to-fines contact) since finer particles had a greater mobility and capacity to easily fit into the internal voids system. Thus, the mechanical role of the main host sand matrix was well maintained and it could consequently provide almost the same overall shear strength, irrespective of the addition of fine particles. Unfortunately,

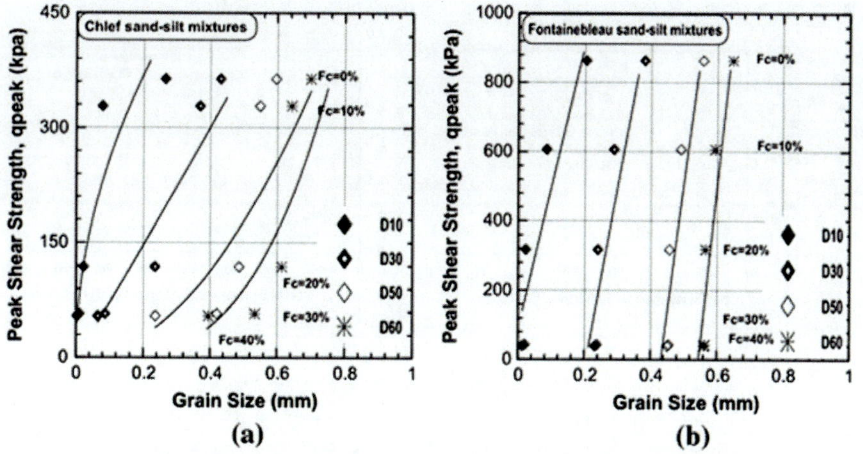

Fig. 1.28 Effect of grain size on the peak shear strength at different fines contents for **a** Chlef sand; **b** Fontainebleau sand (from Cherif Taiba et al. 2016)

the emphasis of the above-mentioned references was on silt grains being the added fines and minor particles such as cohesive clay were not addressed. Consequently, the size range of added fines was comparatively too narrow, rendering an incomplete conclusion about the impact of particle size.

1.4.4 Influence of Gradation Characteristics

Besides particle size, other gradation parameters such as D_{50} of the host sand have also been proved to be an essential factor while studying the undrained shear strength of sand-fines mixtures (Vaid et al. 1990). On the comparison basis of the loosest possible density, static liquefaction resistance of pure sand was found to decrease as sand grains became finer (Monkul et al., 2016). This is in good agreement with Benahmed (2001) and Jradi (2018) who reported that finer Fontainebleau sand ($D_{50} =$ 250 μm) was more sensitive to flow liquefaction than coarser Hostun sand RF (D_{50} = 350 μm) in undrained conditions. Nevertheless, when a clean sand was mixed with fine particles, this trend was interestingly reversed. More explicitly, sand-fines mixtures became more prone to liquefaction as the host sand became coarser.

Based on this controversial observation, Monkul et al. (2016) further concluded that the gap gradation might be a way to understand this troubling issue. The corresponding experimental results showed that the more sand-fines mixture became gap-graded, identified by the increase of uniformity coefficient C_u, the more they became susceptible to liquefaction, directly justified by the apparent decreasing tendency of normalized peak deviator stress q_{peak}/σ'_{3c}, as shown in Fig. 1.29. Later, by conducting direct simple shear tests on silty sand specimens in a wire-reinforced membrane,

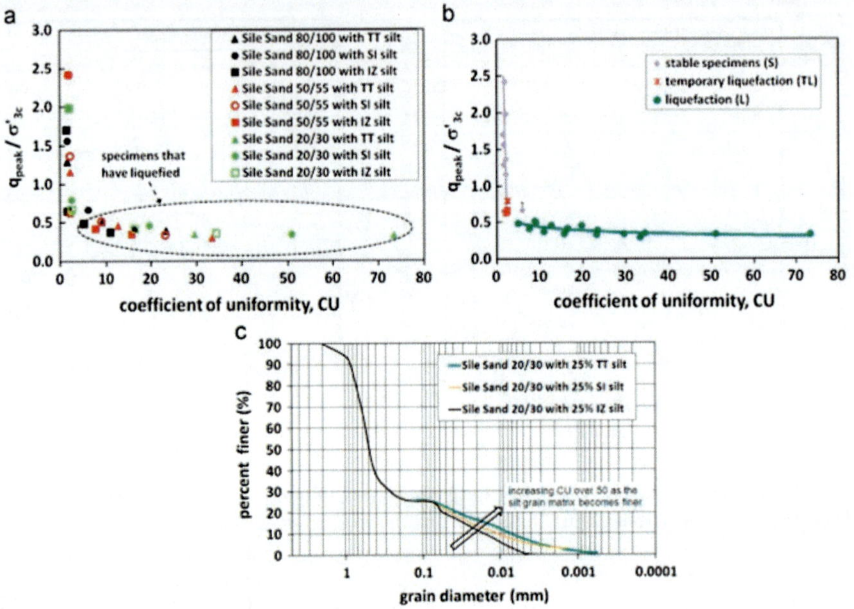

Fig. 1.29 Variation of normalized peak deviator stress $q_{\text{peak}}/\sigma'_{3c}$ with uniformity coefficient C_u (from Monkul et al. 2016)

Doygun et al. (2019) reconfirmed that cyclic shear resistance was further a function of C_u and the decrease in cyclic strength was more pronounced as C_u increased.

1.4.5 Skeleton Void Ratio

It can date back to Mitchell (1976) when the inactive role of fine particles has been discussed while assessing the global structure of coarse matrix grains. Since then, the concept of skeleton void ratio or similar ideas have been extensively employed to interpret the variation in shear resistance of sandy specimens containing fine particles. From a microscopic viewpoint, the fine particles could possibly be accommodated into inter-sand matrix owing to the fact that the size of fine particles is much smaller than that of host sand grains which form the global structure of the soil skeleton. In this sense, fine particles act almost as "voids" and may not necessarily participate in the force chain (Finn et al. 1994) because they are free to move upon shearing. This perspective leads to the definition of skeleton void ratio (Kuerbis and Vaid 1988) in which the volume of those inactive fine particles might be fictitiously assumed as voids:

$$e_s = \frac{e + f_c}{1 - f_c} \tag{1.1}$$

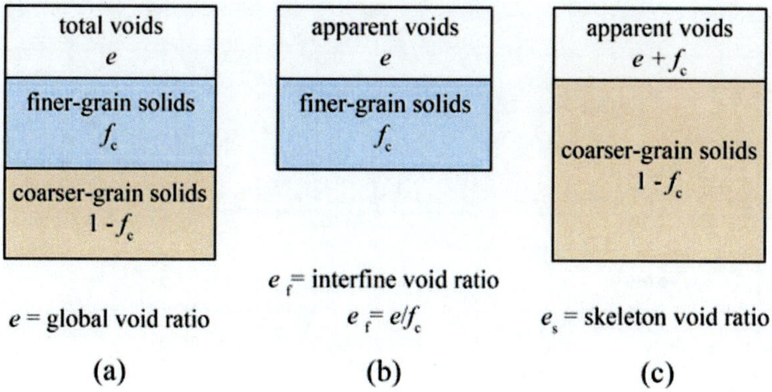

Fig. 1.30 Schematic phase diagrams: **a** silty sand; **b** silt matrix; **c** sand matrix (after Thevanayagam 1998)

where e is the global void ratio and f_c is the fines content. In the published literature, the skeleton void ratio is also referred to as intergranular void ratio (Polito and Martin 2001; Dash and Sitharam 2011) or granular void ratio (Ni et al. 2004). A schematic phase diagram (see Fig. 1.30) was proposed by Thevanayagam (1998) to make clear the difference between global void ratio and skeleton void ratio.

The idea of skeleton void ratio e_s assumes that a composite sandy specimen containing a small amount of fine particles should behave similarly to its host sand forming the global matrix if both are prepared, with due care and diligence, at the same value of e_s. To examine whether or not the above assumption could hold true, a special set of monotonic triaxial tests was conducted by Yang et al. (2015). In Fig. 1.31, the effective stress paths and stress–strain curves are compared for a clean sand specimen (Toyoura sand) and a specimen of the same sand containing 5% silica fines. Although the fines content was controlled to be very low and two specimens were prepared at almost the same e_s (0.950 vs 0.945), two specimens behaved in two distinct ways: the clean sand specimen underwent static liquefaction, whereas the composite sandy specimen developed a strongly dilatant response. Thereby, the presence of fine particles mitigated the liquefaction potential through boosting sand dilatancy if the skeleton void ratio is used as the state variable for comparison. Such an observation is, however, contrary to the key underlying assumptions that fine particles act as voids and make no contribution to global force chain. This logical inconsistency advises that the skeleton void ratio seems not to be a rational state variable for characterizing soil mixtures, even in the case of a small amount of fine particles.

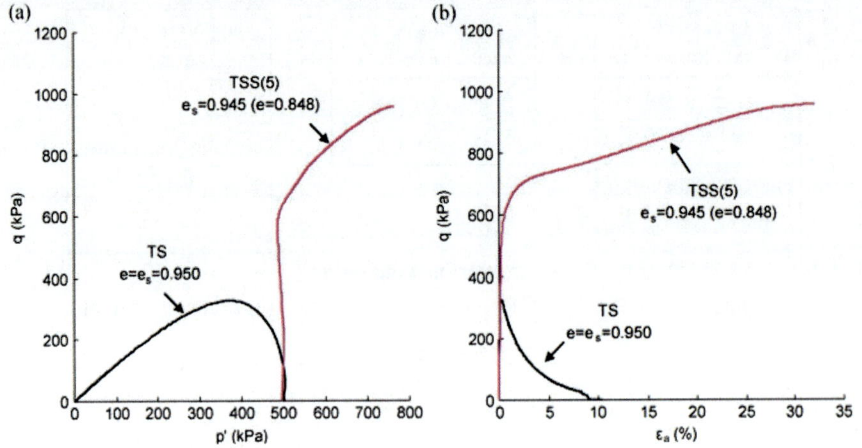

Fig. 1.31 Undrained shear behaviour of Toyoura sand with the addition of fines at the same skeleton void ratio e_s (TS = Toyoura sand and TSS(5) = Toyoura sand mixed with 5% silica fines): **a** effective stress paths; **b** stress–strain curves (after Yang et al. 2015)

1.4.6 Transition Fines Content

With the increase of fines content, the composite soil deposit could possibly pass from one phase to the other through an intrinsic threshold value. At low fines content, sand skeleton is mainly formed and coarse sand grains remain in contact with one another. In that case, fine particles primarily fill the intergranular voids. At higher fines content, sand grains start to disperse in fine particles' matrix and subsequently lose the effective contact with one another. This threshold value distinguishing the sand-predominance from fines-predominance (Karim and Alam 2014) is called "transition fines content" (TFC), "limiting fines content" (FC_{th}) or "critical fines content" (CFC) by different authors. An empirical equation was reported (Hazirbaba 2005) in the literature to roughly estimate TFC using the following expression:

$$\mathrm{TFC} = \frac{W_{\mathrm{fines}}}{W_{\mathrm{sand}} + W_{\mathrm{fines}}} = \frac{G_{\mathrm{f}} \cdot e_{\mathrm{s}}}{G_{\mathrm{f}} \cdot e_{\mathrm{s}} + G_{\mathrm{s}} \cdot (1 + e_{\mathrm{f}})} \tag{1.2}$$

where $W_{\mathrm{fines}}/W_{\mathrm{sand}}$, $G_{\mathrm{f}}/G_{\mathrm{s}}$ and $e_{\mathrm{f}}/e_{\mathrm{s}}$ denote the weight, the specific gravity and the maximum void ratio of fine particles and coarse sand grain, respectively.

Figure 1.33 provides an explicit example demonstrating the manner to experimentally determine de TFC value for a given sand (Hostun sand RF) and fine particles (non-plastic C500 silt) (Gobbi et al. 2022a). As for minimum void ratio e_{min} representing the densest state of composite soil, two distinct phenomena could be identified in the graph: (i) with the initial addition of fine particles, e_{min} decreases owing to the fact that fine particles primarily fill the global sand matrix and (ii) with further addition of fine particles, especially after a threshold value, e_{min} increases owing to the fact that the coarse sand grains start to disperse in the global matrix constructed by fine particles (see Fig. 1.32c). Thus, this threshold value of approximately 20% could be recognized as the value of TFC for these two materials. It is of great importance to note that almost the same value of TFC was arrived by using either the French standard (NFP 94-059 in Fig. 1.33) or the Japanese standard (JGS 0161-2009 in Fig. 1.33).

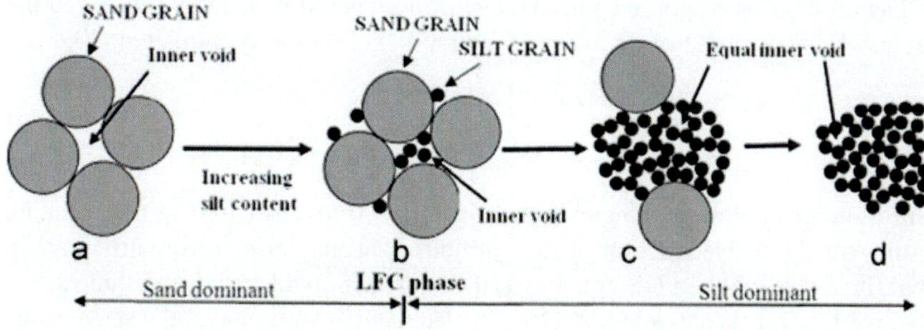

Fig. 1.32 Schematic diagram demonstrating particle arrangement of sand-silt mixture with increased fines content (from Karim and Alam 2014)

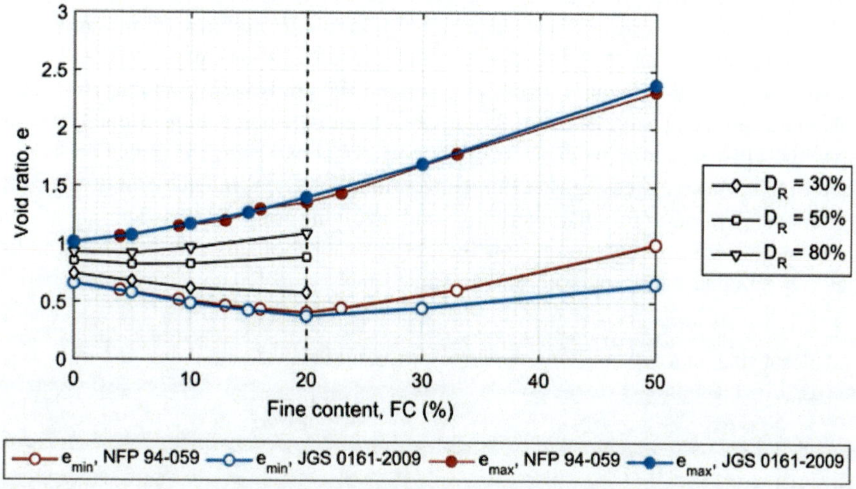

Fig. 1.33 Maximum and minimum void ratio ($e_{\max}$, $e_{\min}$) at different fines contents (Hostun sand RF mixed with non-plastic C500 silt) (from Gobbi et al. 2022a)

1.4.7 Equivalent Void Ratio

While using the skeleton void ratio, it is important to remember the over-idealized behind hypothesis that none of the fine particles is involved in the global force chain. In reality, one should expect that some of the fine particles can possibly be participating in the construction of force chain and transmitting normal/shear stress to their adjacent coarse or fine grains in contact, whilst some others might not. This idea has brought an important improvement for skeleton void ratio and led to the subsequent development of equivalent void ratio e^{eq} (Thevanayagam et al. 2002):

$$e^{\text{eq}} = \frac{e + (1 - b) \times f_c}{1 - (1 - b) \times f_c} \quad (1.3)$$

where a dimensionless parameter b (varying from 0 to 1) is used in the equation standing for the active fraction of fine particles that may effectively participate in global force chain. When b is equal to 1, the equivalent void ratio is the same as the global void ratio e. When b is equal to 0, the equivalent void ratio is the same as the skeleton void ratio. Following up this logic, an enhanced participation of fine particles in the global force chain can be thought by increasing the value of b. In the literature, the equivalent void ratio e^{eq} is also termed as intergranular void ratio (Gobbi et al. 2022a) or equivalent granular void ratio (Rahman et al. 2008; Nguyen et al. 2017). While assessing a series of triaxial results of sand-silt mixtures (Hokksand sand mixed with silt), Yang et al. (2006) proved that a quasi-linear tendency (Fig. 1.34b) existed between the deviator stress at peak and the equivalent granular void ratio (using an empirical value of $b = 0.25$), irrespective of the percentage of fines content up to the TFC. However, no clear correlation was obtained using the global void ratio (see Fig. 1.34a), suggesting that this conventional physical indicator may not be a good state parameter for characterizing sand involving some fine particles.

Despite such a delicate concept, the determination of this dimensionless indicator b remains an open issue and is sometimes subjected to a "trial-and-error"

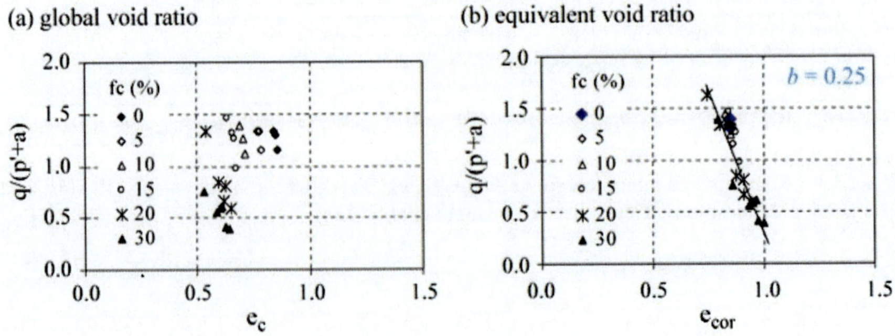

Fig. 1.34 Relationships between stress ratio and **a** global void ratio; **b** intergranular void ratio (f_c from 0 to 30%) (after Yang et al. 2006)

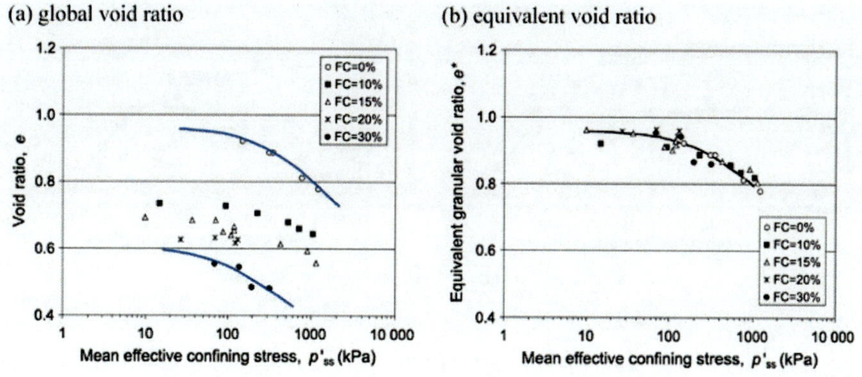

Fig. 1.35 Critical state line for Sydney sand with different fines contents in terms of **a** global void ratio; **b** equivalent void ratio (after Rahman et al. 2008)

back analysis involving many human subjective errors. For simplicity, Rahman et al. (2008) provided a semi-empirical equation based on particle size and fines content as follows:

$$b = \left[1 - \exp(-0.3 \cdot \frac{f_c/f_{\text{thre}}}{k})\right] \times \left(r \cdot \frac{f_c}{f_{\text{thre}}}\right)^r \tag{1.4}$$

where $k = 1 - r^{0.25}$ with $r = d_{50}/D_{10}$ (d_{50}/D_{10} stand for fine/coarse grain equivalent diameter at 50% and 10% finer, respectively) and f_{thre} denotes the transition fines content. With the application of the above formula, an in-depth comparison between the global void ratio and the equivalent void ratio was afterwards made (see Fig. 1.35). In the graph, the remarkable benefit of the equivalent void ratio over the traditional void ratio is its ability to unify the experimental data, irrespective of the fines content (e.g., FC from 0 to 30%) while assessing the critical state line.

More recently, an original formula has been proposed by Gobbi et al. (2022a, 2022b) in order to easily determinate the value of b, which is expressed as follows:

$$b = \frac{e_{\max} - e_{\min}}{1 + e/\text{FC}_{\text{th}}} \tag{1.5}$$

where the numerator ($e_{\max} - e_{\min}$) is indicative of the overall grain-size composition and particle characteristic of a given sand (e.g., size and angularity) as advised by Cubrinovski and Ishihara (2002). By definition, the denominator ($1 + e/\text{FC}_{\text{th}}$) includes both the impacts of current state parameter and transition fines content as previously mentioned. To verify robustness, this equation was applied to a French reference sand HN31 mixed with a silica non-plastic C500 silt at different percentages. Regarding the critical state line in terms of the global void ratio, four distinct curves appear in Fig. 1.36a as a function of fines content. Similar to the conclusion drawn from the above paragraph, this shortcoming can be taken as reasonable since

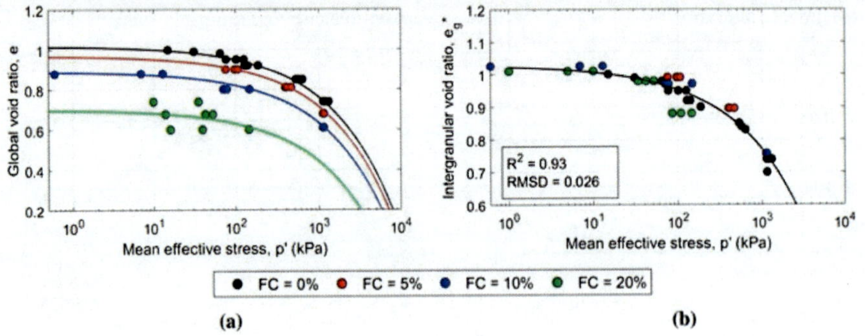

Fig. 1.36 Critical state line for HN31 sand with different percentages of silica C500 non-plastic fines in terms of **a** global void ratio; **b** intergranular void ratio using Eqs. 1.3 and 1.5 (from Gobbi et al. 2022a)

the traditional global void ratio is not capable of distinguishing the mechanical role of fine particles from that of the primary granular material forming the global soil matrix. On the contrary, a unified tendency can be satisfactorily achieved with the use of equivalent void ratio, as directly justified by Fig. 1.36b.

1.4.8 Comparison Between Monotonic and Cyclic Behaviours of Sand-Fines Mixture

In order to synthetically catch on the mechanical behaviours of sand-fines mixture, two comparisons between monotonic and cyclic loadings have been established by Jradi (2018). Fontainebleau sand was taken as the host sand and non-plastic C500 silt was taken as the fine particles.

(i) **Loose state of I_{Dmat} = 0.10 (contractive behaviour)**

Figure 1.37 presents typical results obtained on Fontainebleau sand-C500 mixtures under the consolidation stress of 400 kPa subjected to both monotonic and cyclic loadings. A good consistency can be found between both types of loadings. Regarding the stress–strain curves in Fig. 1.37a, both curves exhibit a complete liquefaction response as q decreases to zero. Similarly, both excess pore water pressure curves in Fig. 1.37b reach the initial consolidation stress equal to 400 kPa. Besides, both types of the effective stress paths are governed by the instability line in Fig. 1.37c.

(ii) **Dense state of I_{Dmat} = 0.50 (dilatant behaviour)**

Figure 1.38 presents the effective stress paths corresponding to the Fontainebleau sand-C500 mixtures under the consolidation stress of 400 kPa subjected to different levels of cyclic shearing. In contrast to the total liquefaction behaviour observed for loose state, the cyclic stress paths do not approach the monotonic stress paths since

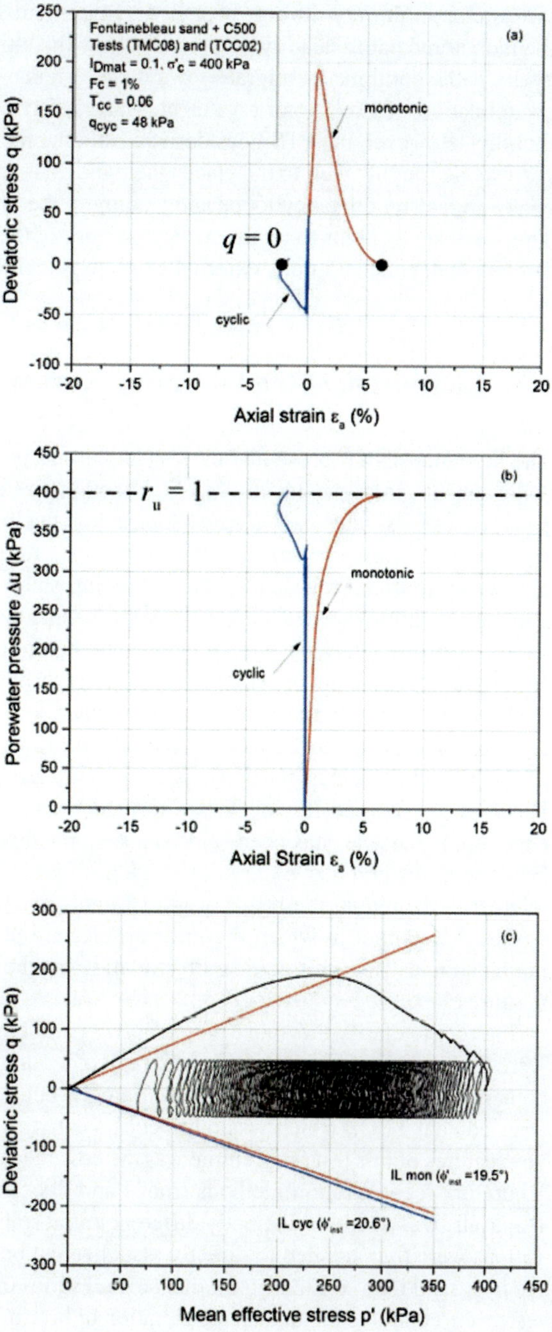

Fig. 1.37 Comparison between monotonic and cyclic behaviour for loose Fontainebleau sand-C500 mixture: **a** stress–strain curves; **b** excess pore water pressure-axial strain curves; **c** effective stress paths (after Jradi 2018)

the latter exhibit very dilatant behaviour. In particular, after the triggering of PTS, the monotonic paths steadily go up and follow the ultimate failure line, whereas the cyclic paths continue to migrate toward the original point because of the gradual accumulation of excess pore water pressure, referring to the occurrence of cyclic mobility. However, the PTS lines derived from the monotonic tests in all three cases correspond roughly well to the place where the dilatant behaviour appears in the first cycles triggering dilatancy loops and leading to the final rupture by cyclic mobility. This consistency helps to retain the key premise of the PTS which is usually viewed as being independent of any external conditions including loading mode.

1.5 Conclusion of the Literature Review

The literature review presented in this chapter has provided a synthesis of the main achievements in the field of sand liquefaction to date. Concerning experimental research, early studies have already focused on the liquefaction of clean sand and a sound understanding has already been developed for these materials. After entering the twenty-first century, sandy soil containing small amounts of fine particles has attracted attention due to a variety of well-documented post-liquefaction field observations. Unfortunately, despite the fact that numerous experimental studies have been successfully carried out, no clear conclusions regarding the effect of fine particles on the liquefaction properties of sands can be drawn so far. The variation of liquefaction resistance has been conventionally studied with some individual physical indicators of fine particles such as plasticity index, particle size, gradation characteristics, etc. However, the composite soil fabric formed by a host sand matrix with the presence of fine particles has been seldom investigated while assessing the liquefaction brittleness. As granular material, the overall resistance of sandy specimens at the microscale is mainly the result of global inter-granular effective contacts. In other words, the manner in which the fine particles are placed within a given host sand matrix should also be a significant factor affecting the global behaviour of sand-fines mixtures.

In order to clarify the above concern, two parallel sets of monotonic triaxial tests were first performed with the consideration of a non-plastic silica powder. The fines content was controlled to be very low (namely, before the attainment of transitional fines content TFC), which was more consistent with field observations. One of the key features of this test programme was the coverage of two common tamping-based laboratory reconstitution methods (moist and dry tamping method) using the same materials so as to create distinctly different initial soil fabrics. Microstructural observations were then applied to directly visualize the binary array of fine particles and the host sand matrix. Finally, another several series of monotonic and cyclic triaxial tests were conducted with the consideration of both non-plastic and plastic fine particles using the dry tamping method. The objective was to identify a representative physical indicator considering the effect of the initial soil fabric.

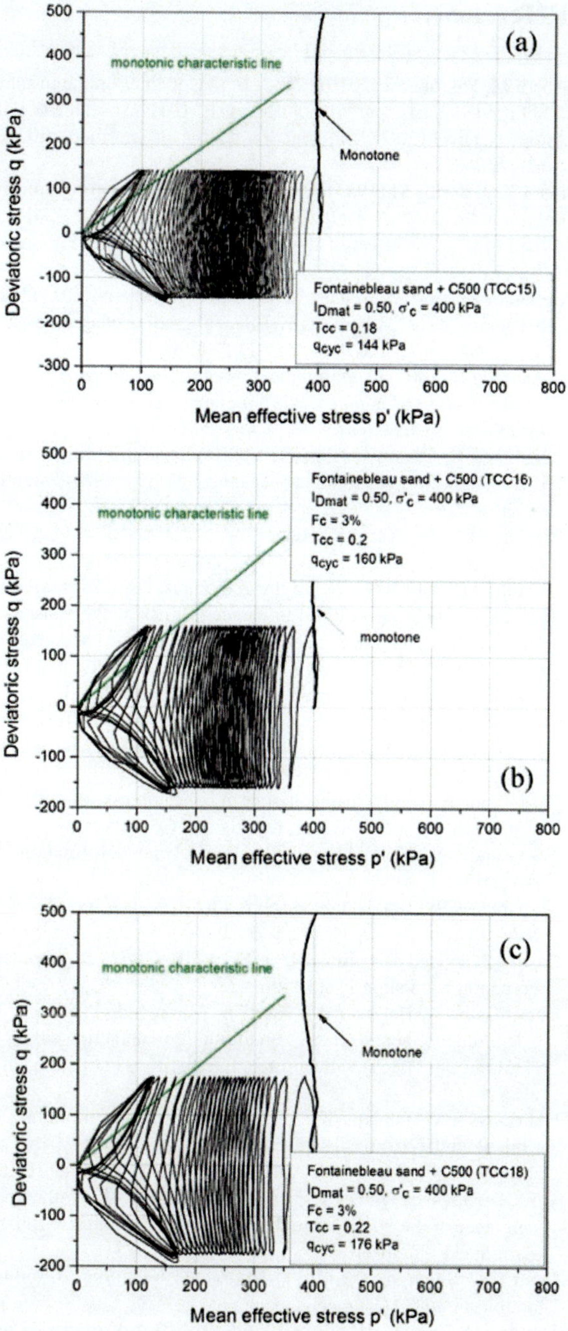

Fig. 1.38 Comparisons between monotonic and cyclic behaviour for dense Fontainebleau sand-C500 mixture in terms of effective stress path with four cyclic stress ratios *CSR* of **a** 0.18; **b** 0.20; **c** 0.22 (after Jradi 2018)

References

Abedi M, Yasrobi SS (2010) Effects of plastic fines on the instability of sand. Soil Dyn Earthq Eng 30(3):61–67. Elsevier. https://doi.org/10.1016/j.soildyn.2009.09.001

Amini F, Qi G (2000) Liquefaction testing of stratified silty sands. J Geotech Geoenviron Eng 126:208–217

Been K, Jefferies M (1985) A state parameter for sands. Geotechnique 35(2):99–112

Belkhatir M, Arab A, Della N, Missoum H, Schanz T (2010) Liquefaction resistance of Chlef river silty sand: effect of low plastic fines and other Parameters. Acta Polytechnica Hungarica 7(2):119–137

Belkhatir M, Arab A, Schanz T, Missoum H, Della N (2011) Laboratory study on the liquefaction resistance of sand-silt mixtures: Effect of grading characteristics. Granular Matter 13(5):599–609. https://doi.org/10.1007/s10035-011-0269-0

Benahmed N (2001) Comportement mécanique d'un sable sous cisaillement monotone et cyclique: application aux phénomènes de liquéfaction et de mobilité cyclique. Ph.D Thesis Ecole Nationale Des Ponts et Chaussées.

Benahmed N, Canou J, Dupla JC (2004) Structure initiale et propriétés de liquéfaction statique d'un sable. Comptes Rendus-Mecanique 332(11):887–894. https://doi.org/10.1016/j.crme.2004.07.009

Benahmed N, Nguyen TK, Hicher PY, Nicolas M (2015) An experimental investigation into the effects of low plastic fines content on the behaviour of sand/silt mixtures. European J Environ Civil Eng 19(1):109–128. Taylor & Francis. https://doi.org/10.1080/19648189.2014.939304

Bouferra R, Shahrour I (2004) Influence of fines on the resistance to liquefaction of a clayey sand. Ground Improvement 8(1):1–5. https://doi.org/10.1680/grim.8.1.1.36366

Canou J (1989) Contribution à l'étude et à l'évaluation des propriétés de liauéfaction d'un sable. Ph.D Thesis Ecole Nationale des Ponts et Chaussées

Carraro J, Prezzi M, Salgado R (2009) Shear strength and stiffness of sands containing plastic or nonplastic fines. J Geotech Geoenviron Eng 135(9, March):1169–1178. https://doi.org/10.1061/(ASCE)1090-0241(2009)135

Casagrande A (1936) Characteristics of cohesionless soils affecting the stability of slopes and earth fills. J Boston Soc Civil Eng, reprinted in Contributions to Soil Mechanics, 257–276

Casagrande A (1976) Liquefaction and cyclic deformation of sands- A critical review. MA, Cambridge

Castro J (1969) Liquefaction of sands. Ph.D Thesis Harvard Soil Mechanics Series n81, Cambridge, MA

Castro BG, Seed RB, Keller TO, Seed HB (1992) Steady-state strength analysis of lower San Fernando Dam slide 118(3):406–427

Cherif Taiba A, Belkhatir M, Kadri A, Mahmoudi Y, Schanz T (2016) Insight into the effect of granulometric characteristics on the static liquefaction susceptibility of silty sand soils. Geotech Geol Eng 34(1):367–382. Springer International Publishing. https://doi.org/10.1007/s10706-015-9951-z

Cubrinovski M, Ishihara K (2002) Maximum and mimimum void ratio characteristics of sands. Soils Found 42(6):65–78

Cubrinovski M, Green RA, Allen J, Ashford S, Bowman E, Brendon B, Cox B, Hutchinson T, Kavazanjian E, Orense R, Pender M, Quigley M, Wotherspoon L (2010) Geotechnical reconnaissance of the 2010 Darfield (Canterbury) earthquake. Bulletin of the New Zealand Society for Earthquake Engineering.

Dash HK, Sitharam TG (2011) Undrained monotonic response of sand-silt mixtures: effect of nonplastic fines. Geomech Geoeng 6(1):47–58. https://doi.org/10.1080/17486021003706796

Doygun O, Brandes HG, Roy TT (2019) Effect of gradation and non-plastic fines on monotonic and cyclic simple shear strength of Silica sand. Geotech Geol Eng 37(4):3221–3240. Springer International Publishing. https://doi.org/10.1007/s10706-019-00838-9

Erten D, Maher MH (1995) Cyclic undrained behavior of silty sand. Soil Dyn Earthq Eng 14(2):115–123. https://doi.org/10.1016/0267-7261(94)00035-F

Finn WD, Ledbetter RH, Wu G (1994) Liquefaction in silty soils: design and analysis. Liquefactioni in silty soils: design and analysis. In: Ground failures under seismic conditions, pp 51–76

Frost JD, Park JY (2003) A critical assessment of the moist tamping technique. Geotech Test J 26(1):57–70. https://doi.org/10.1520/gtj11108j

Ghahremani M, Ghalandarzadeh A, Moradi M (2006) Effect of plastic fines on the undrained behavior of sands. Soil Rock Behav Model, 48–54

Gobbi S, Reiffsteck P, Lenti L, d'Avila MPS, Semblat JF (2022a) Liquefaction triggering in silty sands: effects of non-plastic fines and mixture-packing conditions. Acta Geotechnica 17(2):391–410. Springer, Berlin, Heidelberg. https://doi.org/10.1007/s11440-021-01262-1

Gobbi S, d'Avila MPS, Lenti L, Semblat J-F, Reiffsteck P (2022b) Effect of active plastic fine fraction on undrained behavior of binary granular mixtures. Int J Geomech 22(1). https://doi.org/10.1061/(asce)gm.1943-5622.0002242

Hazirbaba K (2005) Pore pressure generation characteristics of sands and silty sands: a strain approach. Ph.D Thesis University of Texas at Austin

Ishihara K (1995) Soil behaviour in earthquake geotechnics. Oxford University Press

Ishihara K, Yoshimine M (1992) Evaluation of settelements in sand deposits following liquefaction during earthquakes. Soils Found 32(1):173–188. https://doi.org/10.1061/ASCE1090-02412003129:112

Ishihara K, Tatsuoka F, Yasuda S (1975) Undrained deformation and liquefaction of sand under cyclic stresses. Soils Found 15(1):29–44

Ishihara K, Kawase Y, Nakajima M (1980) Liquefaction characteristics of sand deposits at an oil tank site during the 1978 Miyagiken-Oki earthquake. Soils Found 20(2):97–111

Janalizadeh CA, Ghalandarzadeh A, Esmaeili M (2013) Experimental study of the grading characteristic effect on liquefaction resistance of various graded sands and gravelly sands. Arab Geosci 7(7):2739–2748

Jradi L (2018) Study of the influence of fine particles on the properties of liquefaction of sands. Ph.D Thesis University Paris-Est

Jradi L, El Dine BS, Dupla JC, Canou J (2021) Influence of low fines content on the liquefaction resistance of sands. European J Environ Civil Eng, 1–20. Taylor & Francis. https://doi.org/10.1080/19648189.2021.1927195

Karim ME, Alam MJ (2014) Effect of non-plastic silt content on the liquefaction behavior of sand-silt mixture. Soil Dyn Earthq Eng 65:142–150. Elsevier. https://doi.org/10.1016/j.soildyn.2014.06.010

Kolbuszewski JJ (1948) An experimental study of the maximum and minimum porosities of sands. In: Prco 2nd Int Conf. Soil Mechqnics & Foundation Engineering, vol I, pp 168–165

Konrad JM, Flavigny E, Meghachou M (1991) Undrained behavior of loose Hostun sand. Rev Fr Géotech 54:53–63

Kramer SL (1996) Geotechnical Earthquake Engineering

Kramer SL, Seed HB (1988) Initiation of soil liquefaction under static loading conditions. J Geotech Geoenviron Eng 114(4):412–430

Kuerbis R, Vaid Y (1988) Sand sample preparation-the slurry deposition method. Soils Found 28(4):107–118

Ladd RS (1978) Preparing test specimens using undercompaction. Geotech Test J 1(1):16–23

Lade PV (1992) Static instability and liquefaction of loose fine sandy slopes. J Geotech Eng 118(1):51–71. https://doi.org/10.1061/(ASCE)0733-9410(1992)118:1(51)

Lade PV (1993) Initiation of static instability in the submarine Nerlerk Berm. Can Geotech J 30(6):895–904. https://doi.org/10.1139/t93-088

Lade PV, Pradel D (1990) IInstability and plastic flow of soils. I: Experimental observations. ASCE J Eng Mech 116(11):2532–2550

Lade PV, Yamamuro JA (1997) Effects of nonplastic fines on static liquefaction of sands. Can Geotech J 34(6):918–928. https://doi.org/10.1139/t97-052

Levacher D, Garnier J (1994) Reconstitution d'éprouvettes de sable. Appareils de pluviation. Revue Française de Géotechnique 124(12):1215–1217

Liu YJ, Li G, Yin ZY, Dano C, Hicher PY, Xia XH, Wang JH (2014) Influence of grading on the undrained behavior of granular materials. Comptes Rendus-Mecanique 342(2):85–95. Elsevier Masson SAS. https://doi.org/10.1016/j.crme.2013.11.001

Luo G, Zhang JM (2004) Sand constitutive model considering physical state changes. J Hydraul Eng, 26–31

Mitchell JK (1976) Fundamentals of soil behaviour. Wiley, New York

Monkul MM, Yamamuro JA (2011) Influence of silt size and content on liquefaction behavior of sands. Can Geotech J 48(6):931–942. https://doi.org/10.1139/t11-001

Monkul MM, Etminan E, Şenol A (2016) Influence of coefficient of uniformity and base sand gradation on static liquefaction of loose sands with silt. Soil Dyn Earthq Eng 89:185–197. https://doi.org/10.1016/j.soildyn.2016.08.001

Monkul MM, Etminan E, Şenol A (2017) Coupled influence of content, gradation and shape characteristics of silts on static liquefaction of loose silty sands. Soil Dyn Earthq Eng 101:12–26. https://doi.org/10.1016/j.soildyn.2017.06.023

Mulilis J, Seed H, Chan C, Mitchell M, Arulanandan K, Asce M (1977) Effects of sample preparation on sand liquefacton. J Geotech Geoenviron Eng 103(GT2):91–108

Nguyen TK (2014) Etude expérimentale du comportement instable d'un sable silteux. Application aux digues de protection

Nguyen TK, Benahmed N, Hicher PY (2017) Determination of the equivalent intergranular void ratio - Application to the instability and the critical state of silty sand. EPJ Web of Conferences 140:2017–2020. https://doi.org/10.1051/epjconf/201714002019

Ni Q, Tan TS, Dasari GR, Hight DW (2004) Contribution of fines to the compressive strength of mixed soils. Géotechnique 54(9):561–569. https://doi.org/10.1680/geot.54.9.561.56936

Ohmachi T, Tahara T (2011) Nonlinear earthquake response characteristics of a central clay core Rockfill dam. Soils Found 51(2):227–238. https://doi.org/10.3208/sandf.51.227

Ohsaki Y (1970) Effects of sand compaction on liquefaction during the Tokachioki earthquake. Soils Found X(2):112. https://doi.org/10.1061/ASCE1090-02412003129

Papadopoulou AI, Tika TM (2016) The effect of fines plasticity on monotonic undrained shear strength and liquefaction resistance of sands. Soil Dyn Earthq Eng 88:191–206. Elsevier. https://doi.org/10.1016/j.soildyn.2016.04.015

Park SS, Kim YS (2013) Liquefaction resistance of sands containing plastic fines with different plasticity. J Geotech Geoenviron Eng 139(5):825–830. https://doi.org/10.1061/(ASCE)GT.1943-5606.0000806

Phong LM (1980) Phénomènes cycliques dans les sols pulvérulants. Rev Fr Géotech 10(1):39–53

Phong LM, Sidaner JF (1981) Undrained behaviour of cohesionless soils under cyclic and transient loading. In: 1st International Conference on recent advances in geotechnical earthquake engineering and soil dynamics, pp 215–220

Pitman T, Robertson P, Sego D (1994) Influence of fines on the collapse of loose sands. Can Geotech J 31(5):728–739

Polito CP (1999) The effects of non-plastic and plastic fines on the liquefaction of sandy soils. Verginia Polytechnic Institute and State University

Polito PC, Martin RJ (2001) Effects of nonplastic fines on the liquefaction resistance of sands. J Geotech Geoenviron Eng 127(5):408–415

Rahman MM, Lo SR, Gnanendran CT (2008) On equivalent granular void ratio and steady state behaviour of loose sand with fines. Can Geotech J 45(10):1439–1456. https://doi.org/10.1139/T08-064

Roscoe K, Schofield A (1963) Mechanical behaviour of an idealized wet clay. In: Proceddedings of the European Conference on Soil Mechanicas and Geotechnical. Wiesbaden, Germany, pp 47–54

Seed HB, Idriss IM, Arango I (1983) Evaluation of liquefaction potential using field performance data. J Geotech Eng 109(3):458–482

Seed HB, Tokimatsu K, Hader LF, Chung RM (1985) Influence of SPT procedures in soil liquefaction resistance evaluations. J Geotech Eng 111(12):1425–1445

Shengcong F, Tatsuoka F (1984) Soil liquefaction during Haicheng and Tangshan earthquake in China; a review. Soils Found 24(4):22–29

Stamatopoulos CA (2010) An experimental study of the liquefaction strength of silty sands in terms of the state parameter. Soil Dynam Earthquake Eng 30(8):662–678. Elsevier. https://doi.org/10.1016/j.soildyn.2010.02.008

Thevanayagam S (1998) Effect of fines and confining stress on undrained shear strength of silty sands. J Geotech Geoenviron Eng 124(6):479–491. https://doi.org/10.1061/(ASCE)1090-0241(1999)125:11(1024)

Thevanayagam S, Shenthan T, Mohan S, Liang J (2002) Undrained fragility of clean sands, silty sands, and sandy silts. J Geotech Geoenviron Eng 128(10):849–859. https://doi.org/10.1061/jggefk

Tokimatsu K, Yoshimi Y (1983) Empirical correlation of soil liquefaction based on SPT N-value and fines content. Soils Found 23(4)

Tsukamoto Y, Ishihara K (2022) Advances in soil liquefaction engineering. Springer

Vaid YP, Fisher JM, Kuerbis R, Negussey D (1990) Particle gradation and liquefaction. J Geotech Eng 116(4):698–703

Verdugo R (1992) Characterisation of sandy soil behavior under large deformation. University of Tokyo

Xenaki VC, Athanasopoulos GA (2003) Liquefaction resistance of sand-silt mixtures: An experimental investigation of the effect of fines. Soil Dyn Earthq Eng 23(3):1–12. https://doi.org/10.1016/S0267-7261(02)00210-5

Yang J, Wei LM (2012) Collapse of loose sand with the addition of fines: The role of particle shape. Geotechnique 62(12):1111–1125. https://doi.org/10.1680/geot.11.P.062

Yang SL, Sandven R, Grande L (2006) Instability of sand-silt mixtures. Soil Dyn Earthq Eng 26(2–4 SPEC. ISS.):183–190. https://doi.org/10.1016/j.soildyn.2004.11.027

Yang J, Wei LM, Dai BB (2015) State variables for silty sands: global void ratio or skeleton void ratio? Soils Found 55(1):99–111. Elsevier. https://doi.org/10.1016/j.sandf.2014.12.008

Yoshida N, Tazoh T, Wakamatsu K, Yasuda S, Towhata I, Nakazawa H, Kiku H (2007) Causes of Showa bridge collapse in the 1964 Niigata earthquake based on eyewitness testimony. Soils Found 47(6):1075–1087. https://doi.org/10.3208/sandf.47.1075

Yoshimine M, Ishihara K (1998) Flow potentail of sand during liquefaction. Soils Found 38(3):189–198

Zhu Z, Cheng W (2020) Parameter evaluation of exponential-form critical state line of a state-dependent sand constitutive model. Appl Sci 10(1):328. https://doi.org/10.3390/app10010328

Zhu Z, Kham M, Alves Fernandes V, Lopez-Caballero F (2020) Dynamic response of a Central Clay Core dam under two-component seismic loading. Springer Series in Geomech Geoeng 1:231–236. https://doi.org/10.1007/978-3-030-46351-9_23

Zhu Z, Zhang F, Dupla J, Canou J, Foerster E (2021a) Assessment of tamping-based specimen preparation methods on static liquefaction of loose silty sand. Soil Dyn Earthq Eng 143(January):106592. Elsevier Ltd. https://doi.org/10.1016/j.soildyn.2021.106592

Zhu Z, Zhang F, Peng Q, Chabot B, Dupla J (2021b) Development of an auto compensation system in cyclic triaxial apparatus for liquefaction analysis. Soil Dyn Earthq Eng 144(December 2020):106707. Elsevier Ltd. https://doi.org/10.1016/j.soildyn.2021.106707

Chapter 2
Tested Materials, Equipment and Protocols

2.1 Tested Materials

2.1.1 Hostun Sand HN31

The clean sand used in this thesis is a poorly graded French reference sand called Hostun sand HN31, which is characterized by uniform sub-angular grains and widely used in many laboratories in Europe. The grain size distribution curve of HN31 sand determined by the laser diffraction method is shown in Fig. 2.1. The principle of the method is that while passing through a laser beam, a particle could scatter the light at a certain angle and intensity linking to its equivalent grain size. The detailed index properties of this sand are listed in Table 2.1.

2.1.2 Fine Particles

The fine particles selected in the present work are respectively: (i) C500 (a non-plastic silica powder), (ii) Arvel illite (a plastic clay of illite mineralogical content) and (iii) Speswhite (a plastic kaolinite clay). The relevant physical properties of these particles are given in Table 2.2. The specific gravity G_s is found to be the same, equal to 2.65.

The plasticity indexes of Arvel illite and Speswhite kaolinite were determined following the French Standard NF P 94-051 (see Fig. 2.2a). As for C500 silica power, it should be noted that it was challenging to obtain a standard roller determining the corresponding plasticity value although various attempts had been made to follow the same standard. Thereby, this type of fine particles is logically assumed to be non-plastic (NP) fines owing to a plasticity index $I_p < 10$.

For simplicity, the grain size distribution curve of non-plastic C500 silica powder was also determined by the laser diffraction method (see Fig. 2.3), the same as HN31 sand in the above section. However, the grain size distribution curve of the cohesive

Z. Zhu, *Influence of Fine Particles on the Liquefaction Properties of a Reference Sand*, Springer Theses, https://doi.org/10.1007/978-3-031-24299-1_2

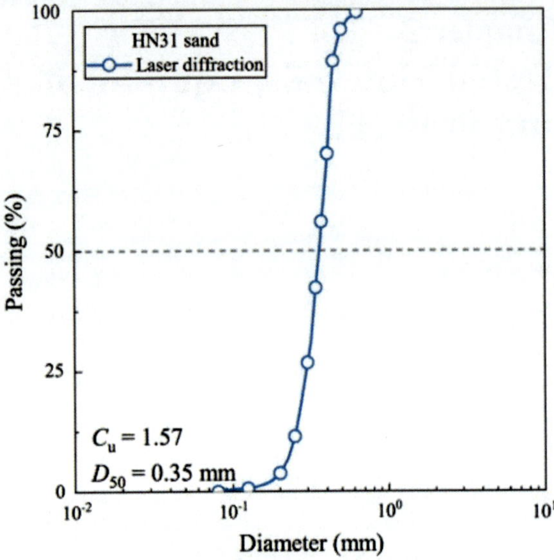

Fig. 2.1 Grain size distribution curve of HN31 Hostun sand

Table 2.1 Physical properties of HN31 sand

Material	D_{50} (μm)	C_u	e_{min}	e_{max}	G_s
HN31	350	1.57	0.656	1	2.65

Table 2.2 Physical properties of fine particles

Material	d_{50} (μm)	Mineralogical content	G_s	I_p
C500	6.0	silica	2.65	NP
Arvel illite	2.2	illite		34
Speswhite kaolinite	0.7	kaolinite		30

plastic soils such as Arvel illite and Speswhite kaolinite were differently measured, using the sedimentation method (see Fig. 2.2b) since the method is more accurate and more precise for finer soils, especially for particles having mean diameter d_{50} values smaller than 80 μm. The principle of the method is to measure the sedimentation rate of soil particles in a static liquid (water) which is then related to the grain size distribution of the particles.

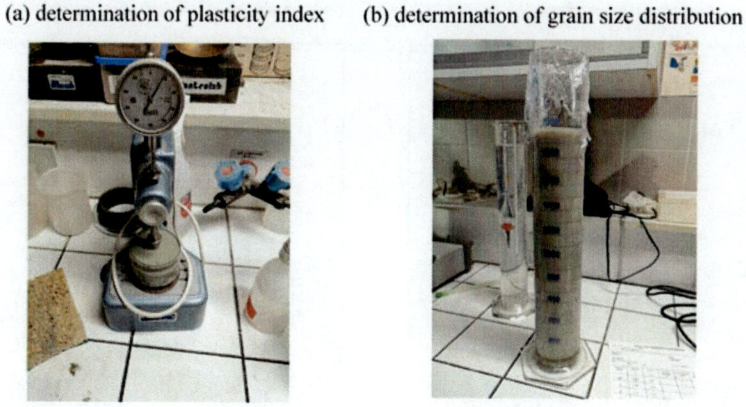

Fig. 2.2 Views showing the determination of: **a** plasticity index (fall cone); **b** grain size distribution curve (sedimentometry)

Fig. 2.3 Grain size distribution curves for the fine particles selected

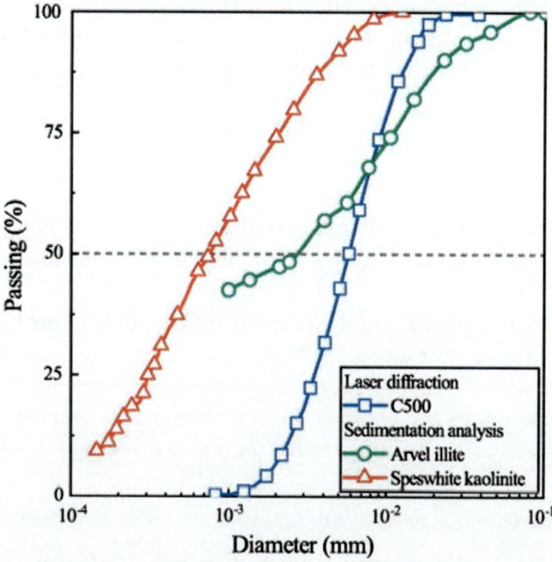

2.2 Specimen Preparation and Reconstitution Methods

2.2.1 *Definitions*

Several well-documented field observations suggested that, at real liquefied sites, the fine particles commonly exist in a limited percentage. Thereby, one might always expect that the in situ liquefiable soils are generally composed of a coarse sand matrix but polluted, to some extent, by a very small quantity of fine particles. Based on this

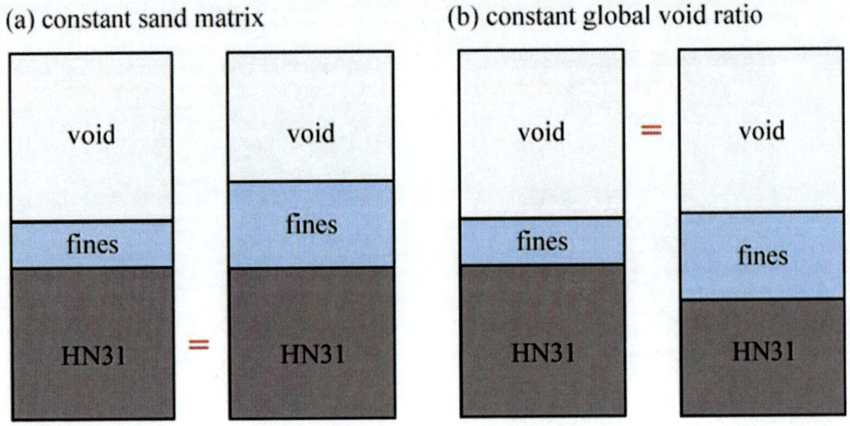

Fig. 2.4 Concept of: **a** constant sand matrix; **b** constant global void ratio

configuration, the void ratio of ***sand matrix*** e_{mat} instead of the global void ratio (see Fig. 2.4) was considered as the controlled parameter while preparing test specimens through adopting the density index of sand matrix I_{Dmat} as follows:

$$I_{Dmat} = (e_{max} - e_{mat})/(e_{max} - e_{min}) \tag{2.1}$$

where e_{mat} is the void ratio of ***sand matrix***, and where e_{max} and e_{min} represent the maximum and minimum void ratio of ***clean*** HN31 sand, respectively. The clear benefit of this concept is the ability to keep the quantity of coarse sand forming the global soil matrix constant, irrespective of the fines content (see Fig. 2.4a). And the fines content F_c in the present work is then defined as the dry mass of added fine particles over that of the host sand matrix. Here, it is important to mention that there is no difference with respect to the algebraic value of I_{Dmat} between the clean sand specimens and composite sandy specimens composed of the same quantity of sand matrix since the added fines are viewed as empty void.

2.2.2 *Specimen Reconstitution Methods*

In this study, two tamping-based reconstitution methods were employed to prepare uniform soil specimens for triaxial testing (with a diameter and height of 100 mm and 200 mm, respectively): (i) dry tamping (DT) and (ii) moist tamping method (MT). For the dry tamping method, pure sand and added fines were firstly mixed by stirring and shaking in a container (see Fig. 2.5). Afterwards, all mixtures were divided into 10 equal parts and each part was carefully introduced into a split mould with a spoon, followed by slight compaction efforts to achieve the required thickness (20 mm) using a hand tamper. As for loose composite specimens, particular attention should

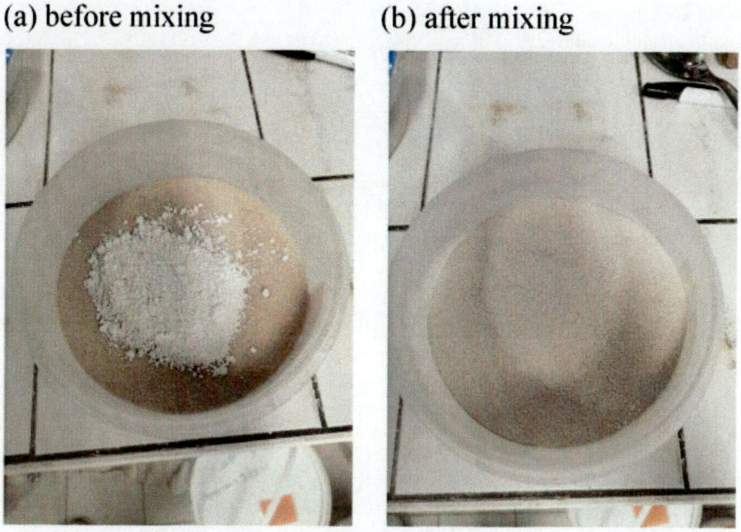

Fig. 2.5 Views of sand-fines mixture: **a** before mixing; **b** after mixing (case of speswhite)

be paid to minimizing the drop height in order to (i) reduce the potential energy as much as possible to meet the desired density and (ii) avoid any possible segregation between coarse and fine particles during freefall.

Moist tamping has been widely used in many experimental investigations related to liquefaction since it is readily able to create very loose specimens with the presence of capillary force, which is always viewed as beneficial particularly in liquefaction analysis, where extremely contractive responses are desirable. In this research, the technique consisted of mixing sand and fine particles in a large blender in wet condition created by adding 5% water (by mass) to the dry mass of sand. Similar to the above dry tamping method, the mixtures were then divided into 10 identical parts and each part was consecutively poured into a membrane-lined split mould (see Fig. 2.6) and compacted to 20 mm before the next layer was set in place. Note that, while keeping the sand matrix unchanged with the gradual addition of fine particles, the fragile latex membrane did not allow very powerful compaction force. Therefore, only low fines contents F_c (smaller or equal to 10%) were reachable with this method.

2.3 Equipment and Protocols Used for Monotonic Loading

2.3.1 Experimental Setup

The triaxial device used for monotonic testing is shown in Fig. 2.7. The vertical displacement rate of the electromechanical press is configured through the controlling panel (maximum velocity equal to 10 mm/min). The confining chamber allows supporting the maximum confining pressures of 2 MPa. This capacity meets most of

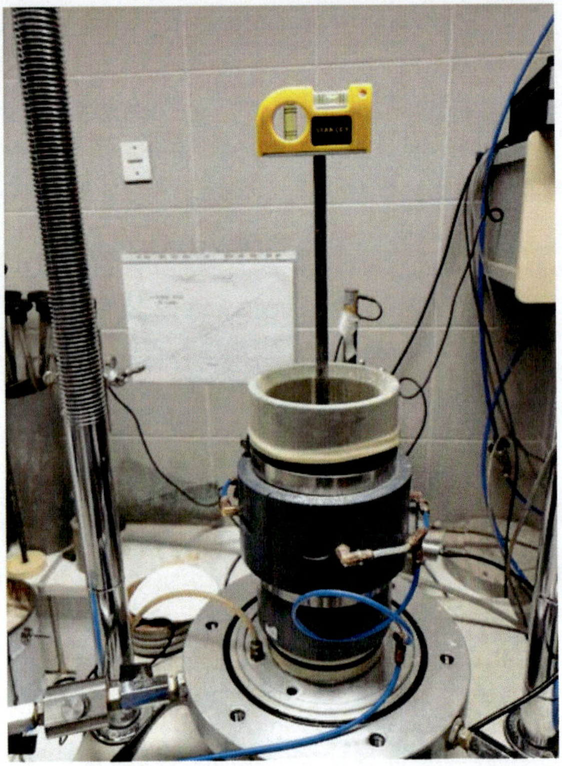

Fig. 2.6 View of the membrane-lined split mould during specimen reconstitution process

the experimental requirements for sand liquefaction analysis, especially for studying the liquefaction properties of shallow sandy deposits. Two force sensors are used. The first one is located outside the confining chamber, recording the force variation during the test combined with ram friction (between the loading piston and confining chamber). The second one is located inside the chamber and allows the measurement of the pure force variation on the test specimen without ram friction. Therefore, the value recorded by the internal force sensor is considered for further processing.

2.3.2 *Testing Protocols*

Once a specimen (100 mm in diameter and 200 mm in height, as shown in Fig. 2.8) has been reconstituted by one of the above reconstitution methods, the top triaxial base is mounted and carefully sealed with 2 water-tight O-rings. Subsequently, a slight vacuum of about 25 kPa is directly applied inside the test specimen to minimize any possible disturbances while removing the split mould, as well as to check possible leakage of the latex membrane.

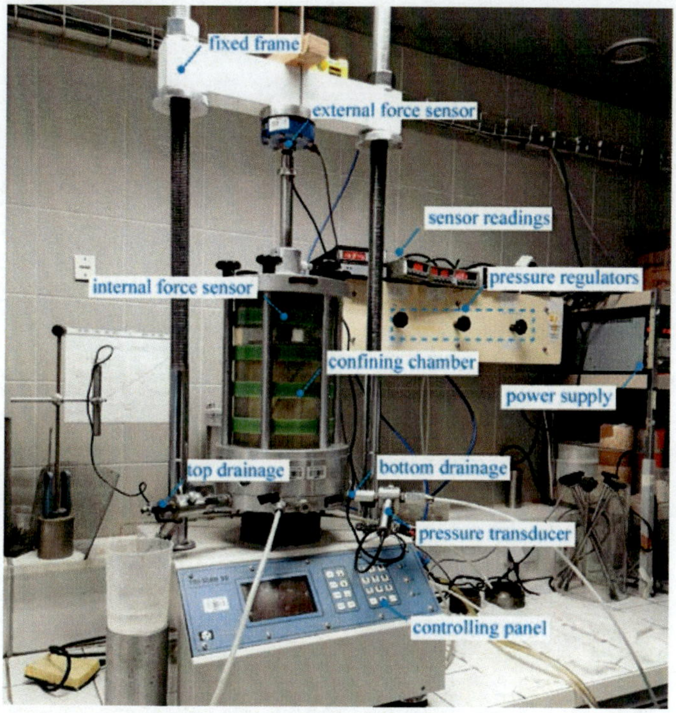

Fig. 2.7 General view of the triaxial setup used for monotonic testing

Before initiating the three-step saturation, the cell pressure is increased to 100 kPa to maintain the specimen properly. Carbon dioxide (CO_2) is firstly flushed (15 kPa) through the dry specimen for about 15 min with the purpose of replacing air, followed by the injection of pre-made de-aired water (9 kPa). The key action during this operation is the fact that the flux of both CO_2 and de-aired water should be seriously limited being very light to avoid disturbing the fragile soil fabric containing fine particles. The final step of saturation involves the application of a back pressure: the back pressure and confining pressure are alternatively increased in small steps of about 20 kPa until the Skempthon's B value ($B = \Delta u/\Delta p_{\text{conf}}$, the variation of excess pore water pressure Δu divided by the confining pressure increment Δp_{conf}) is at least equal to or greater than 0.98. This criterion is believed to be fair enough to achieve a satisfactory saturation for sandy specimens. All monotonic triaxial tests presented in this work were performed at a constant axial strain rate of 0.50% per minute. Subjected to monotonic shear, this strain rate is thought to be slow enough to allow the equilibrium of the pore water pressure throughout the test specimen. During the shearing process, this aspect is re-confirmed by the pore water pressure readings at both bottom and top drainage lines that are directly connected to the test specimen, respectively. It should be noted that all monotonic triaxial tests were

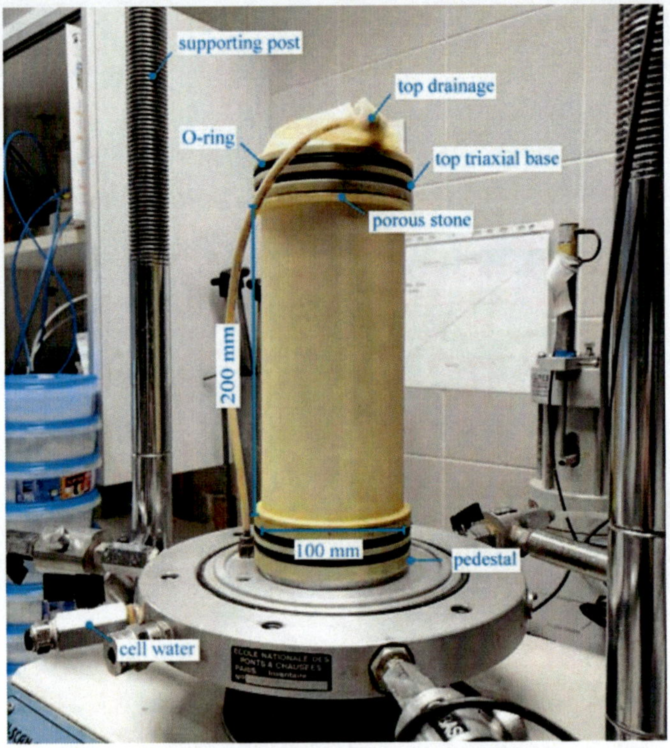

Fig. 2.8 View of a reconstituted specimen after removing the split mould (with a vacuum of 25 kPa) on the monotonic triaxial apparatus

carried out at a controlled ambient temperature of 20 ± 1 °C, avoiding the need to consider any unwanted thermo-coupling phenomenon.

2.3.3 Validation Tests with Fine Particles

Provided that the plastic fines are highly suspectable to stick together in the presence of water while using the moist tamping method, a series of monotonic triaxial tests (VM1, VM2 and VM3) was then performed on HN31 sand-Speswhite mixtures to check the repeatability level of the test. All three specimens were prepared with the same testing parameters ($I_{\mathrm{Dmat}} = 0.00$, $\sigma'_{\mathrm{c}} = 100$ kPa and $F_{\mathrm{c}} = 1.0\%$). The subplots of Fig. 2.9 present the stress–strain curves, excess pore water pressure curves and effective stress paths, respectively. The stress–strain curves in Fig. 2.9 were all characterized by a clear drop after the initial rapid growth, referring to the occurrence of static liquefaction. This fact is consistent with the evolution of excess pore water pressure depicted in Fig. 2.9b. Besides, all stress paths in Fig. 2.9c were governed by the instability line. Nevertheless, each test specimen developed its own response upon shearing, rendering the credibility of the reconstitution procedure to be very low. Thus, the test specimens containing ***plastic fine particles*** (namely, Speswhite kaolinite and Arvel illite) were solely reconstituted using the dry tamping method in this thesis. The corresponding repeatability tests will be provided in the following chapters. Recently, the effect of mixing orders on the microstructure of sand-clay mixtures has been examined by Yin et al. (2021). Three orders via altering the moment to introduce three materials (dry Fsand N34, dry K Clay and distilled water) were created and several analytical methods (scanning electron microscopy, X-ray, etc.) were used to visualize the corresponding microstructures. For all specimens, plastic clays were found to heterogeneously disperse in micrometre-sized layers surrounding host sand grains, conforming to the above test results.

In order to straightforwardly check the homogeneity of specimens containing non-plastic fines (C500) while using the moist tamping method, a preliminary test ($I_{\text{Dmat}} = 0.00$, $\sigma'_{\text{c}} = 100$ kPa and $F_{\text{c}} = 2.5\%$) was performed adopting the protocol for monotonic loading as aforementioned. Before initiating the triaxial setup, a small soil

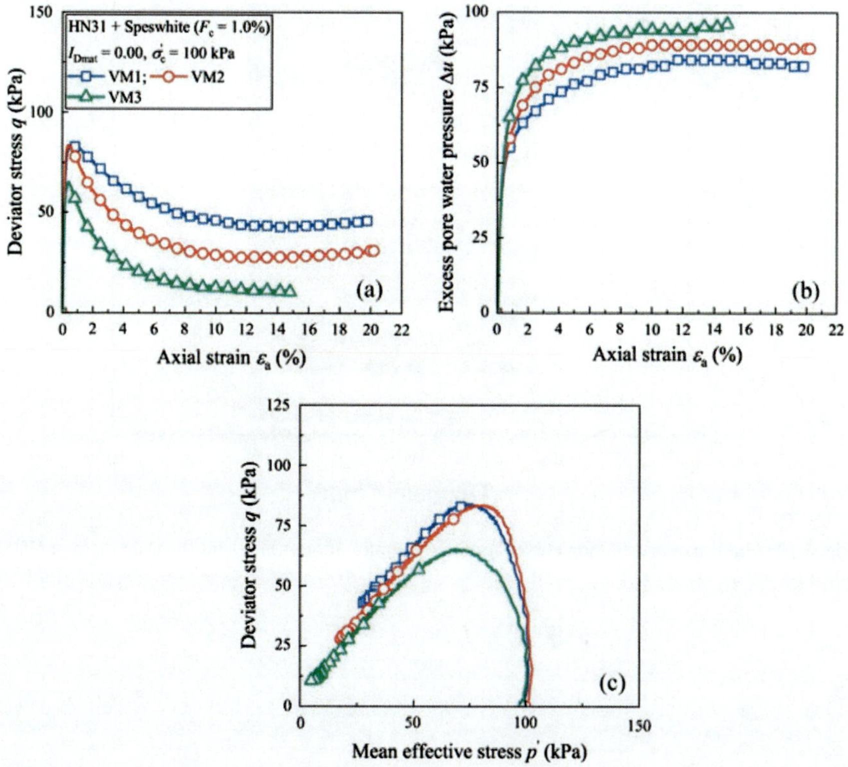

Fig. 2.9 Validation tests on HN31-speswhite mixtures: **a** stress–strain curves; **b** excess pore water pressure curves; **c** effective stress paths

sample (HN31 sand + C500) was directly taken as soon as the soil was successfully mixed in the blender. After all triaxial procedures, a slight vacuum of about 10 kPa was supplied through the bottom drainage line to keep the test specimen undisturbed, as well as to avoid the unloading collapse. Finally, the cell water was slowly evacuated and the confining chamber was removed. Another two small soil samples were thereafter taken with a spoon from two positions within the rubber membrane: (i) one from the upper part and (ii) another one from the middle part around the shear band. Figure 2.10 presents the corresponding grain size distribution curves of these samples. In the graph, three curves were almost the same, indicating that the protocol adopted can be considered to be fairly appropriate to meet the requirement of the soil homogeneity throughout all experimental procedures.

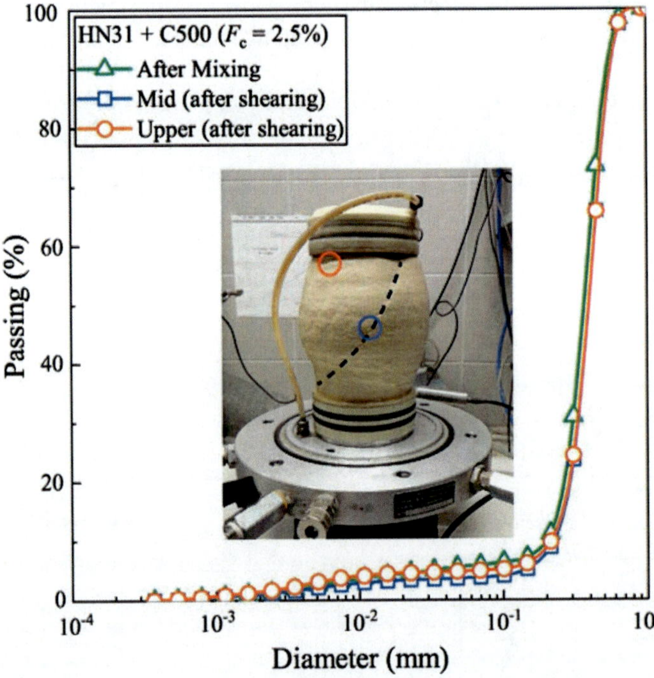

Fig. 2.10 Grain size distribution curves for three samples after mixing and shearing (case of C500 silica powder)

2.4 Equipment and Protocols Used for Cyclic Loading

2.4.1 Experimental Setup

The complete setup for cyclic testing is mainly composed of the following components: (i) a reaction frame equipped with a 50 kN servohydraulic actuator (MTS system), (ii) a triaxial cell, (iii) an interactive software, (iv) CO_2 supply source and (v) a de-airing water system. A schematic view presenting the essential reaction frame and the confining chamber is presented in Fig. 2.11a. The reaction frame consists of two supporting columns and a rigid slab. The rigid slab is horizontally embedded by 10 strong screws into the reaction frame for supporting the confining chamber after the specimen is set up. All elements of the reaction frame were fabricated to be significantly rigid, which eliminated the need to consider any undesirable deformation of the reaction frame itself rather than that of the test specimen. Controlling, loading and measurement are automatically accomplished with the aid of a user-friendly interactive commercial software "MTS FlexTest 60".

In order to generate the accurate loading signal with high robustness either in displacement- or force-controlled mode, the hydraulic actuator is regulated by a closed-loop PID(F) (Proportional-Integral-Derivative) controller. The axial displacement and force are measured by a LVDT (Linear Variable Differential Transformer, with a precision of about 0.01 mm) and a force sensor (with a precision of about 0.05 kN) that is located under the pedestal of the specimen for the purpose of capturing

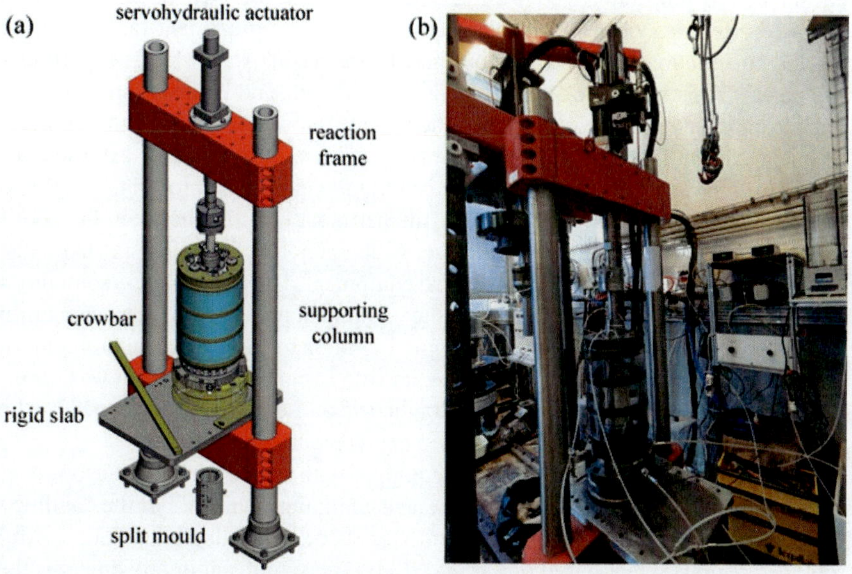

Fig. 2.11 **a** 3D drawing of the servohydraulic triaxial setup; **b** general view

the pure value without any parasite ram friction. It should be mentioned that the force and displacement sensors could easily be both zeroed to directly record the deviator force and the pure axial displacement upon shearing. Besides, the cell pressure is measured using a pressure transducer with an accuracy of about 1 kPa. During all experimental stages, an oscilloscope can be launched to display the axial variation either in terms of displacement or force, allowing the relative error in between to be examined in real-time.

In order to reverse the loading direction from triaxial compression to extension (or vice versa), most current cyclic triaxial configurations require an initial connection between the loading piston and the test specimen, leading, more or less, to the introduction of a pre-shearing effect on the specimen. In that case, the test specimen cannot steadily achieve an isotropic consolidation state with $K_c = 1$ during the consolidation as well as the saturation stages with the increase in cell and back pressures since there is always a part of the upper triaxial base has been covered by the loading piston, as shown in Fig. 2.12a. A common solution is to force the test specimen to sustain the same axial force by using the coupled vertical actuator in order to add the missing force maintaining null the load cell measurement (see Fig. 2.12b). The missing force is equivalent to the variation of cell pressure (Δp_{cell}) multiplied by ***the piston area A_p***. However, either compressive or extensional pre-shearing could possibly occur in a repeating manner due to the following two independent factors: (i) the axial compensation (through the coupled vertical actuator) could not be done at exactly the same time with the increase of cell pressure (e.g., through a pressure–volume controller or pressure regulator) and a minor time lag between these two actions always exists; (ii) the reaction of the vertical actuator is never instant and always governed by the sensible closed-loop tuning controlling parameters and the type of actuator. This undesired effect has been repeatedly recognized to be a prominent factor altering the sand liquefaction response. For instance, Ishihara and Okada (1982) experimentally revealed that when pre-sheared in a static manner, sand liquefaction resistance depended not only on the magnitude of the applied pre-shearing force but also more conspicuously on the direction of pre-shearing (either compressive or extensional). As a consequence, the test results obtained with the above test procedure might be questionable. In the published literature, the pertinent coupling between the above two components remains a doubtful issue, which is not well detailed in standards, papers, and even in manuals of commercial equipment. Thus, researchers sometimes develop their own procedures based on the equipment and their experiences. From the academic and practical viewpoints, a good solution to these practical problems in cyclic triaxial test is necessary. Therefore, an auto compensation system was developed and incorporated into the servohydraulic triaxial setup used (developed by the geotechnical team of Navier laboratory) to overcome this issue.

The triaxial cell is composed of a base equipped with a force transducer, a top cover equipped with a waterproof piston guide and a confining chamber. For the confining chamber, the latest vacuum-based system was successfully ungraded (Fig. 2.13b) from the previous screw-based one (Fig. 2.13a). The significant improvement is the use of (i) a cylindrical metallic part attached to the actuator with a central hole

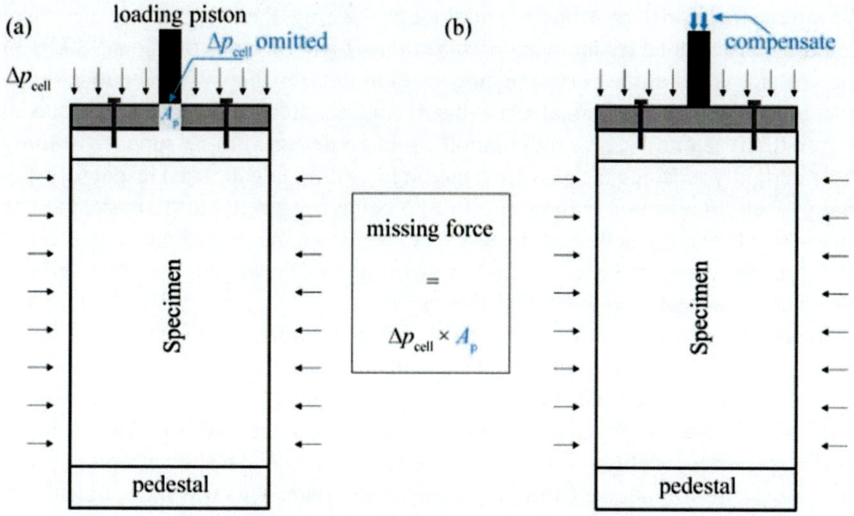

Fig. 2.12 Schematic diagrams showing: **a** the connection problem between the loading piston and the upper triaxial base; **b** the usual solution through the axial compensation

connecting to a vacuum tube and (ii) a special upper base with a flexible plastic ring (marked in orange in Fig. 2.13b).

The main objective of these two components is to guarantee that during the consolidation and saturation stages, there is no physical contact between the ram of the actuator and the test specimen, achieving a true isotropic consolidation state with

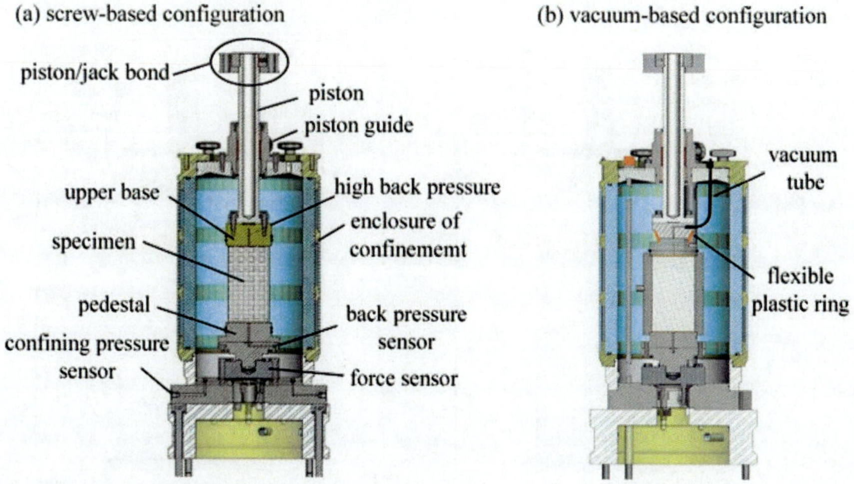

Fig. 2.13 2D simplified cross-sections of the triaxial cell showing: **a** the previous screw-based configuration; **b** the advanced vacuum-based configuration

$K_c = 1$, as mentioned previously. And prior to shearing, the coupling between them could be easily done by applying a ***continuous*** high vacuum (of about 98 kPa in the present work) in the corresponding vacuum tube. It should be mentioned that, as compared with conventional screw-based configuration, the special upper triaxial base in the vacuum-based configuration could be designed to be super light (***only 3 N***), certainly reducing the undesired inertia force affecting the sand response, especially when liquefaction collapse is close to being triggered. Furthermore, there is no torsional shearing acting on the test specimen since the final connection is made with the high vacuum instead of rudely twisting the screws into the upper triaxial base that is connected to the test specimen.

While attempting to make the physical contact between the piston and the upper triaxial base through imposing a high vacuum, the vacuum tube alone is not capable of maintaining the required isotropic consolidation state. Once an enclosed space is formed by the loading piston and the flexible plastic ring, the further movement of the loading piston inevitably compresses the inside air layer and results in a high excess air pressure destroying the isotropic consolidation state. This was indeed identified in a preliminary trial that an excess air force of more than 250 N was recorded when the cell pressure was equal to 300 kPa, although the loading piston was very slowly lowered down to approach the upper triaxial base using the displacement-controlled mode. This fact provides a persuasive reason to design a novel auto compensation system in order to well control this kind of error. To solve this technical problem, the proposed auto compensation system consists of coupling the cell pressure and vacuum tubes together through a single one to an air pressure regulator on which two valves were respectively installed, as shown in Fig. 2.14b.

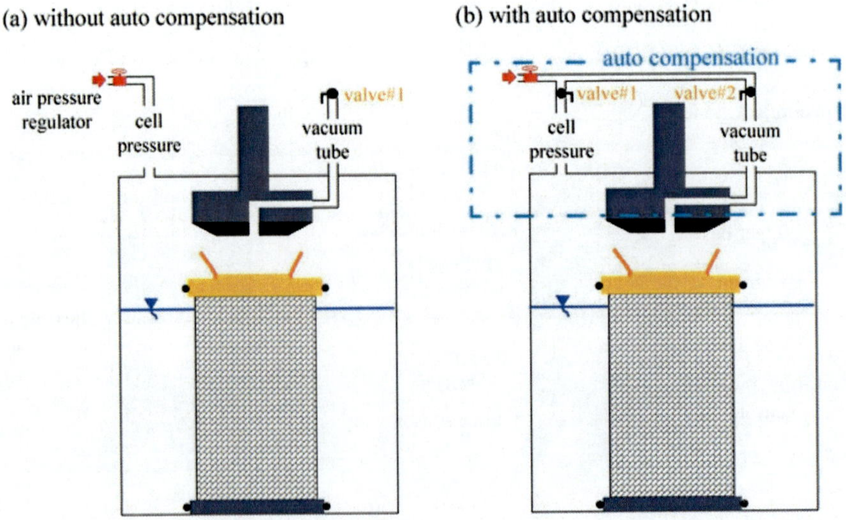

Fig. 2.14 Schematic drawings of the confining chamber: **a** without auto compensation system; **b** with auto compensation system coupling cell pressure tube and vacuum tube with two valves

While lowering down the loading piston with this auto compensation system, two valves are completely open. In this sense, any unwanted excess air pressure could be automatically equilibrated by the connected air pressure regulator so as to hold the isotropic consolidation state. Before applying the final high vacuum in the vacuum tube, it is necessary, in the first place, to release the inside consolidation pressure. With the auto compensation system, the removal of this pressure can be done in a secure manner, instead of suddenly truing on the valve on the vacuum tube.

2.4.2 Testing Protocols

Hereafter, the detailed cyclic triaxial test procedure is described, especially with the use of the auto compensation system. Once a triaxial specimen (100 mm in diameter and 200 mm in height) has been formed by either one of the two reconstitution methods (dry tamping or moist tamping method), the flexible plastic ring is carefully put on the central part of the upper triaxial base for the future connection. In particular, the horizontality of the upper base (see Fig. 2.15a) needs to be seriously checked by a spirit level. A minor gap (see Fig. 2.15b) should be ensured at this stage while installing the loading piston through an overhead crane. The purpose of this minor gap is to achieve a true isotropic consolidation state during the saturation and consolidation stages. After the confining chamber is set up, the auto compensation system can be activated (by opening both valve#1 and valve#2) and the confining pressure in the chamber is then increased to 100 kPa in order to well hold the test specimen. Finally, the conventional three-step saturation (involving CO_2, de-aired water and back pressure) could be started, the same as that presented in the monotonic triaxial test programme. After saturation, the connection process can be initiated. The axial force and displacement sensors are both zeroed with the purpose of clearly detecting possible variations. The loading piston was slowly lowered in the ***displacement-controlled mode*** in order to touch the upper triaxial base till having a small contact force value of about 90 N. In this process, the use of auto compensation system avoids the occurrence of any confusing excess air pressure acting on the test specimen. In general, as long as a true mechanical contact between the loading piston and the upper triaxial base is achieved, a sudden release of air pressure in the vacuum tube would not lead to any record variation in load cell measurement since the vertical movement of the test specimen is totally constrained by the loading piston. However, in a preliminary test, a significant extensional force of about − 300 N (rebound phenomenon) was indeed recorded by the oscilloscope.

This could be attributed to the fact that although various efforts had been made to level the surface of the test specimen, an absolute horizontal one could be only seldom achieved, and a minor inclination always existed. This argument explains the necessity to check the horizontality of the test specimen at the beginning stage. In order to overcome this inherent shortcoming, ***force control*** is used in the controlling system. And a force value slightly (of about 100 N) higher than that occurred in the contact moment (of about 90 N) is required in the system so as to maintain the mechanical

Fig. 2.15 Views showing: **a** checking horizontality of the upper base; **b** adjustment of the loading piston (a minor gap is kept between the loading piston and the upper triaxial base); **c** coupled cell pressure and vacuum tubes through the auto compensation system

contact properly between the loading piston and the upper triaxial base and avoid the rebound phenomenon as explained above. Regarding the auto compensation system at this time, the vacuum tube is isolated through closing valve#1 (valve#2 still open). Under the circumstances, the air pressure inside the vacuum tube can be removed in a secure manner. During this process, the loading piston maintaining 100 N (***in force-controlled mode***) would automatically correct the horizontality of the test specimen with quasi-static movement (generally smaller than 0.1 mm). Although the cell pressure tube is closed, the cell pressure in the confining chamber remains still unchanged since this minor movement of the loading piston cannot possibly lead to observable air pressure variation in such a voluminous air–liquid system. It should be noted that low sensitive PID tuning parameters are here used to avoid the hypercorrection phenomenon, which is capable of exceptionally compressing the test specimen.

Finally, it is necessary to sum up the possible disturbance that can appear in the aforementioned cyclic triaxial test programme: (i) the maximum additional axial force in sand history that the test specimen can undergo is about ***100 N***, only slightly higher than the force value in the contact moment, which is inevitable in any cyclic triaxial system and (ii) the maximum additional axial displacement is about 0.1 mm (equal to 0.05% in axial strain) while removing the inside confining pressure in the vacuum tube; whereas the majority of this item is the ***horizontality correction*** of the upper triaxial base instead of the test specimen itself. As for macroscopic triaxial tests with this minor disturbance, the test specimen can be logically supposed to be "quasi-intact" in a "pseudo-elastic" range.

2.4.3 Validation Tests

To assess the performance of the complete triaxial equipment with the new auto compensation system, two validation tests owing exactly the same testing parameters were performed with the latest vacuum-based configuration and auto compensation system. For further analysis, a different specimen was also performed with the conventional screw-based configuration. As compared with the vacuum-based configuration, the most different point while using the screw-based configuration was that after removing the split mould, the loading piston and the upper triaxial base were immediately coupled by inserting strong screws. Afterwards, any increase in cell pressure was carried out in ***small steps*** of about 20 kPa, followed by the manual compensation (of about 110 N in each incremental step) with the use of the actuator. All three specimens were prepared at the same density state of $I_{Dmat} = 0.70$ and subjected to the same level of shearing (cyclic stress ratio $CSR = \tau/\sigma'_c = 0.22$). The detailed validation test programme is summarized in Table 2.3.

The evolutions of imposed deviator stress σ_d, excess pore water pressure Δu and axial strain ε_a in terms of number of cycles N_{cyc} of two tests (VC1 versus VC2) are shown in subplots of Fig. 2.16. It can be seen that all curves almost coincided with each other, exhibiting satisfactory repeatability for the adopted experimental protocols involving the use of auto compensation system. Regarding Δu in Fig. 2.16a, the "double peak mechanism" became increasingly apparent after N_{cyc} equal to 15,

Table 2.3 Characteristics of the validation tests performed

Reference	I_{Dmat}	σ'_c (kPa)	Reconstitution method	CSR	System	Compensation
VC1	0.70	200	DT	0.22	*Vacuum*	Auto
VC2	0.70	200	DT	0.22	*Vacuum*	Auto
VC3	0.70	200	DT	0.22	*Screw*	Manual

revealing the outset of cyclic mobility. Regarding ε_a in Fig. 2.16c, two curves were almost overlapping in the initial loading stage (see local zoom in the graph), whilst some small differences were indeed observed at large axial strains. This is because, near the failure stage, the test specimens developed a "fluid" potential. As a result, the actuator could not be capable of keeping up with this transient displacement in real-time although an advanced mechanical controller was set. This phenomenon can also be directly embodied by the "distorted" sinusoidal shape, as underlined in Fig. 2.16a.

Figure 2.17 presents the two corresponding effective stress paths in which a high confidence level of repeatability is found. Both stress paths were characterized by the "Butterfly wing", reconfirming the triggering of cyclic mobility. Near the end of the test, in extension half cycles, the deviator stress q exceeded the prescribed value (−88 kPa), which was attributed to the increasingly important extensional deformation, rendering the effective area of the test specimens much smaller.

The test results obtained with the vacuum-based configuration (in blue) and with the screw-based configuration (in orange) are plotted together in Fig. 2.18. For the screw-based configuration, there was an abrupt downward drifting of ε_a (as outlined in the black circle), whilst the evolution of ε_a with the vacuum-based configuration was almost flat and slowly tilted to the extension side. Similar observations can also be found in Fig. 2.18b that the liquefaction potential was significantly boosted

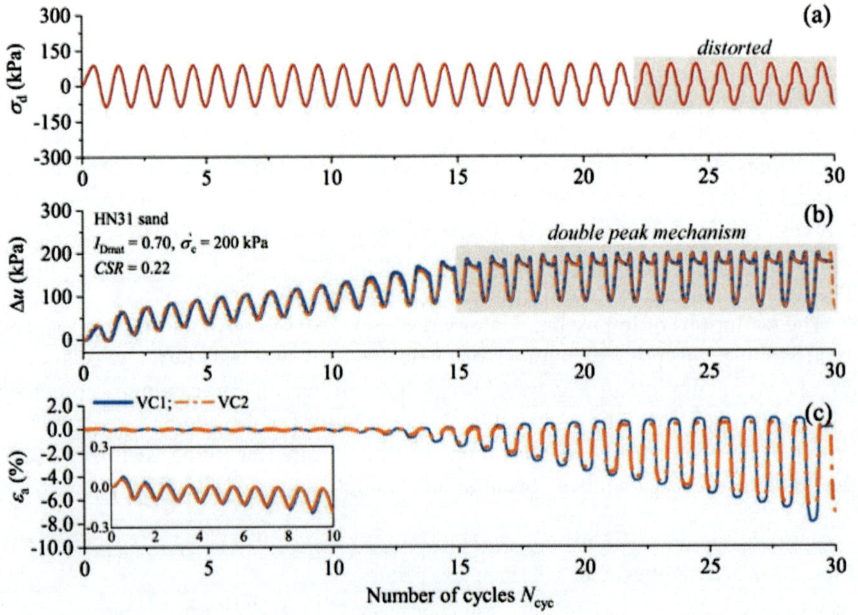

Fig. 2.16 Comparison of two repeatability tests with $I_{Dmat} = 0.70$, $\sigma'_c = 200$ kPa and CSR = 0.22: **a** imposed deviator stress; **b** excess pore water pressure versus number of cycles; **c** axial strain versus number of cycles

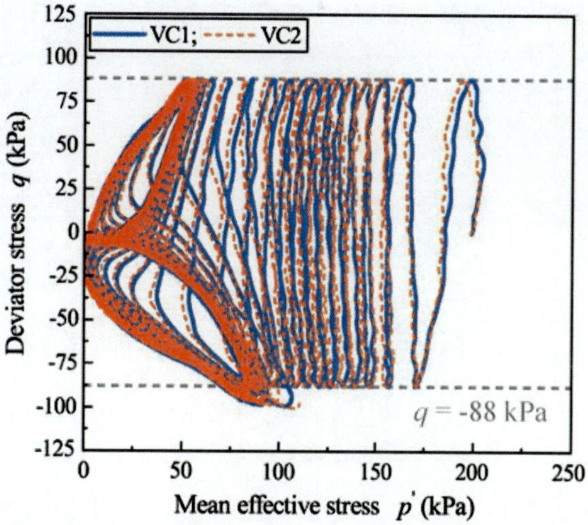

Fig. 2.17 Comparison of two repeatability tests with $I_{Dmat} = 0.70$, $\sigma'_c = 200$ kPa and CSR = 0.22 in terms of effective stress path

while using the screw-based configuration where a manual compensation was always needed.

The above distinctly different observations can be only possibly ascribed to the initial fabric anisotropy, caused by the pre-shearing effect in sand history associated with different triaxial configurations since all testing parameters such as density, consolidation stress were exactly kept the same. When sand specimen is deposited under gravity, it has been experimentally revealed (Oda 1972; Yang et al. 2008b) that the long axes of sand grains preferentially orient along the horizontal direction achieving a naturally stable state, as conceptually shown in Fig. 2.19a. As for screw-based configuration, during the consolidation step as well as the saturation step

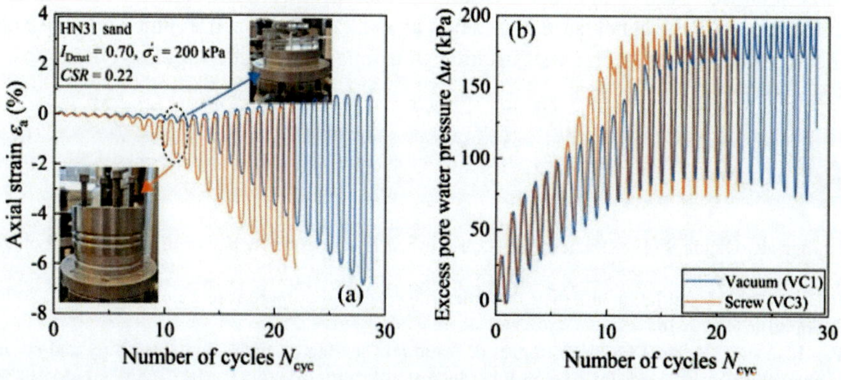

Fig. 2.18 Comparison between vacuum and screw-based configuration with $I_{Dmat} = 0.70$, $\sigma'_c = 200$ kPa and CSR = 0.22: **a** axial strain versus number of cycles; **b** excess pore water pressure versus number of cycles

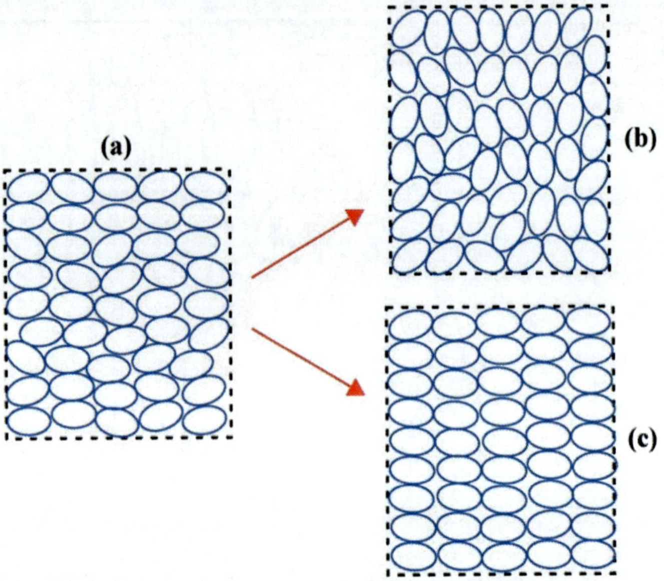

Fig. 2.19 Schematic diagrams showing **a** initial soil fabric; **b** effect of extensional pre-shearing; **c** effect of compressive pre-shearing

with the application of back pressure, there is always a repeating extensional pre-shearing acting on the test specimen since the manual compensation of axial force through the vertical actuator could not be synchronously realized with the increment in pressure. Despite the fact that there is no significant volume change that occurs in these two steps while undergoing the above extensional pre-shearing, the soil fabric becomes gradually weakened since more particles tend to lie in the unstable vertical direction (see Fig. 2.19b) and subsequently more prone to sand liquefaction. This is consistent with the conclusion drawn by Pan et al. (2018) that undrained response of medium-dense sand decreases as static extensional stress ratio ($q_{\text{pre}}/\sigma'_{\text{c}}$) increases.

References

Ishihara K, Okada S (1982) Effects of large preshearing on cyclic behavior of sand. Soils Found 22(3):110–125

Oda M (1972) Inital fabric and their relations to mechanical properties of granular material. Soils Found 12(1):17–36

Pan K, Yang ZX, Xu TT (2018) Impact of static preshearing on undrained anisotropy and shear characteristics of sand. Int J Geomech 18(12):1–12. https://doi.org/10.1061/(ASCE)GM.1943-5622.0001319

Yang ZX, Li XS, Yang J (2008) Quantifying and modelling fabric anisotropy of granular soils. Geotechnique 58(4):237–248. https://doi.org/10.1680/geot.2008.58.4.237

Yin K, Fauchille AL, Di Filippo E, Othmani K, Branchu S, Sciarra G, Kotronis P (2021) The influence of mixing orders on the microstructure of artificially prepared sand-clay mixtures. Adv Mater Sci Eng. https://doi.org/10.1155/2021/8552224

Chapter 3
Behaviour Under Monotonic Loading

This chapter presents the behaviour of clean sand and sand-fines mixtures under undrained monotonic triaxial testing. As far as the moist tamping method is concerned, only non-plastic C500 silica particles were involved, owing to the homogeneity concern. Based on the test results obtained, special attention was paid to verifying the impact of fine particles on the variation of the instability line (loose state), the phase transformation state PTS line (medium-dense state) and the critical state line. As for the dry tamping method, all three fine particles (plastic/non-plastic) were adopted. Owing to the fact that the HN31-C500 specimens reconstituted with two methods behaved totally in an opposite manner, microstructural observations were thereafter performed to explore the possible influence of initial soil fabric, as well as to propose a reasonable mechanism behind. Finally, a physical indicator, on the basis of particle size, is proposed to summarize the effect of fine particles.

3.1 Experimental Testing Programme

Regarding the moist tamping method with non-plastic C500 silica, two density indexes ($I_{\mathrm{Dmat}} = 0.00$ for loose state and 0.50 for medium-dense state) were chosen. And two different consolidation stresses were used ($\sigma'_c = 100$ and 400 kPa) to verify the unification of different concepts of soil behaviour. Regarding the dry tamping method with the consideration of all three fine particles (plastic and non-plastic fines), the specimens were prepared under the same test conditions ($I_{\mathrm{Dmat}} = 0.00$ and $\sigma'_c = 100$ kPa) so as to emphasize the impact of fine particles themselves on the undrained behaviour under monotonic loading. The detailed experimental programme is summarized in Table 3.1.

Z. Zhu, *Influence of Fine Particles on the Liquefaction Properties of a Reference Sand*, Springer Theses, https://doi.org/10.1007/978-3-031-24299-1_3

Table 3.1 Summary of the monotonic triaxial test programme

Reference	Type of fines	F_c (%)	I_{Dmat}	σ'_c (kPa)	Reconstitution method
MTT01	–	0.0	0.00	100	MT (1)
MTT02	–	0.0	0.00	100	MT
MTT03	C500	1.0	0.00	100	MT
MTT04	C500	2.5	0.00	100	MT
MTT05	C500	5.0	0.00	100	MT
MTT06	C500	10.0	0.00	100	MT
MTT07	C500	0.0	0.00	400	MT
MTT08	C500	1.0	0.00	400	MT
MTT09	C500	2.5	0.00	400	MT
MTT10	C500	5.0	0.00	400	MT
MTT11	C500	10.0	0.00	400	MT
MTT12	C500	0.0	0.50	100	MT
MTT13	C500	1.0	0.50	100	MT
MTT14	C500	2.5	0.50	100	MT
MTT15	C500	5.0	0.50	100	MT
MTT16	C500	10.0	0.50	100	MT
MTT17	C500	0.0	0.50	400	MT
MTT18	C500	1.0	0.50	400	MT
MTT19	C500	2.5	0.50	400	MT
MTT20	C500	5.0	0.50	400	MT
MTT21	C500	10.0	0.50	400	MT
MTT22	C500	5.0	0.00	100	DT (2)
MTT23	C500	10.0	0.00	100	DT
MTT24	C500	10.0	0.00	100	DT
MTT25	C500	15.0	0.00	100	DT
MTT26	Illite	5.0	0.00	100	DT
MTT27	Illite	10.0	0.00	100	DT
MTT28	Illite	15.0	0.00	100	DT
MTT29	Speswhite	5.0	0.00	100	DT
MTT30	Speswhite	10.0	0.00	100	DT
MTT31	Speswhite	10.0	0.00	100	DT
MTT32	Speswhite	15.0	0.00	100	DT

Note (1) Moist tamping, (2) Dry tamping

3.2 Typical Test Results

In this section, typical test results are given to describe the main responses of clean sand and sand-fines specimens. The assessment of repeatability is a very important item to check the reliability of the adopted reconstitution methods, as well as to validate both the experimental setup and the corresponding protocol. With this in mind, the typical test results shown below are systematically accompanied by the corresponding repeatability test.

3.2.1 Clean Sand Behaviour

Figure 3.1 shows the results of typical static liquefaction tests conducted on two very loose specimens of clean HN31 sand reconstituted with the moist tamping method for $I_{\text{Dmat}} = 0.00$ and $\sigma'_{\text{c}} = 100$ kPa. The stress–strain curves in Fig. 3.1 were evidently characterized by a remarkable drop after the initial growth in deviator stress q. As shown in Fig. 3.1b, the corresponding excess pore water pressure Δu was generated in a rapid manner without any drop tendency and finally achieved an ultimate value very close to the consolidation stress of 100 kPa, referring to pure contractancy. Regarding the effective stress paths, after passing through a deviator stress value at peak, the two paths migrated toward the original point, indicating a complete liquefaction collapse.

Although some minor differences indeed exist in Fig. 3.1, the two tests can still be believed to be consistent with each other. This brings confidence to the experimental method adopted in the present work.

3.2.2 Sand-Fines Behaviour

Figure 3.2 presents the results of typical tests conducted on sand-fines mixtures (dry tamping method) with the addition of two types of fine particles: (i) non-plastic C500 at the left and (ii) plastic Speswhite at the right. For the stress–strain curves in Fig. 3.2a, a dilatant behaviour was observed for HN31-C500 specimens, whilst a limited liquefaction behaviour was found for HN31-Speswhite specimens. The same conclusion can also be found with respect to the evolution of excess pore

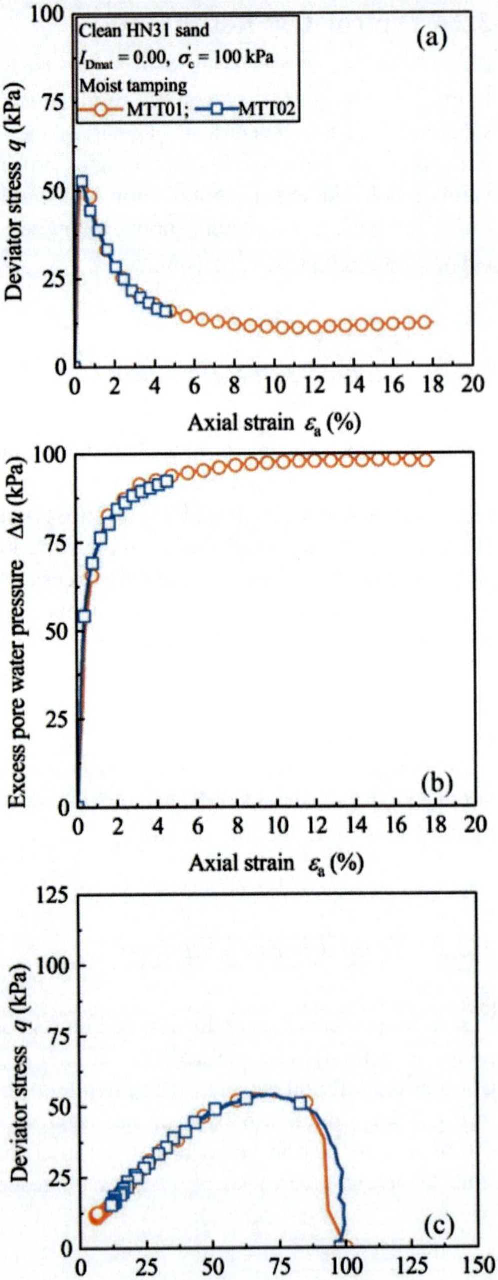

Fig. 3.1 Results obtained for clean HN31 sand specimens with I_{Dmat} = 0.00, σ'_c = 100 kPa: **a** stress–strain curves; **b** excess pore water pressure curves; **c** effective stress paths

water pressure in Fig. 3.2b. The effective stress paths for HN31-C500 specimens were governed by the PTS line, whilst those for HN31-Speswhite specimens were controlled by the instability line. In addition, a high level of repeatability was verified again.

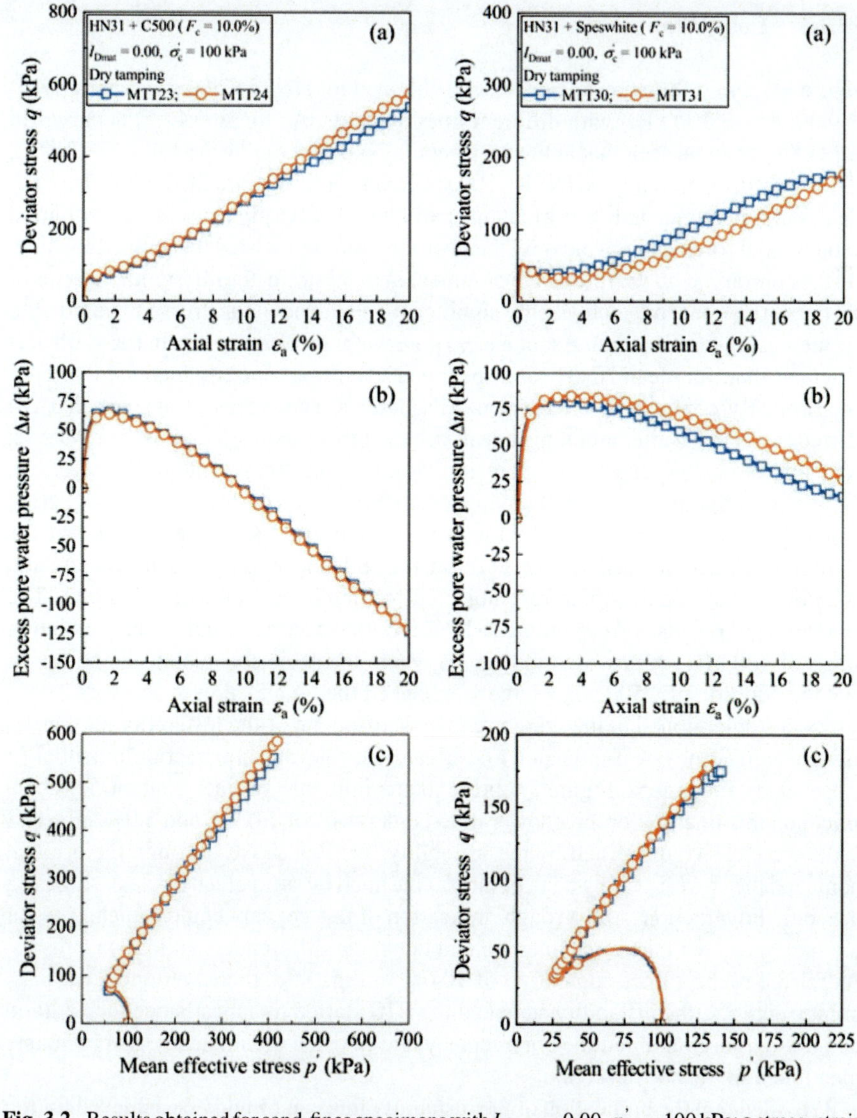

Fig. 3.2 Results obtained for sand-fines specimens with $I_{\text{Dmat}} = 0.00$, $\sigma'_c = 100$ kPa: **a** stress–strain curves; **b** excess pore water pressure curves; **c** effective stress paths

3.3 Influence of Fine Particles

3.3.1 Influence of C500 Silica Using the Moist Tamping Method

3.3.1.1 Loose State

Figure 3.3 shows the experimental results obtained for HN31-C500 specimens (I_{Dmat} = 0.00, σ'_{c} = 100 kPa) with different fines contents. As for stress–strain curves in Fig. 3.3a, it can be seen that at the very beginning of loading (especially $\varepsilon_{\mathrm{a}} < 0.5\%$), all curves corresponding to HN31-C500 specimens almost coincided with that of the clean sand specimen, indicating that fine particles did not truly make any contribution to the global force chain. And only the host sand matrix resisted the applied loading. On the contrary, with the further development of axial strain, the difference in terms of deviator stress became increasingly significant and the deviator stress increased with the increasing fines content. As for excess pore water pressure curves in Fig. 3.3b, it is observed that, for clean HN31 sand specimen, Δu monotonously increased without any drop. By contrast, in the cases of HN31-C500 specimens, it appears that Δu started to decrease after attaining a peak value, corresponding to a dilatant response. And this effect became much more important as the fines content F_{c} increased. As for the effective stress paths depicted in Fig. 3.3c, the peak value triggering flow liquefaction increased to be much more important as F_{c} increased. All the above phenomena suggest that the addition of C500 indeed mitigated the liquefaction potential for sand-fines mixtures subjected to undrained monotonic loading. The experimental results corresponding to HN31-C500 specimens under a consolidation stress of 400 kPa are provided in Fig. 3.4 where the same conclusion can be drawn that the addition of C500 silica largely enhanced the liquefaction resistance.

As aforementioned in the literature review, the slope of the instability line serves as an adequate physical indicator to evaluate the relevant liquefaction potential of loose sandy specimens. Figure 3.5 presents the influence of fines content F_{c} (C500 silica) on this line. As for three lower fines contents F_{c} of 1.0, 2.5 and 5.0%, all stress paths were clearly governed by the instability line that passes through the original point and the corresponding peak points (as outlined by solid green circles) triggering the static liquefaction. And a rough unification of this concept could be achieved for the specimens with the same fines content but under different consolidation stresses. As for the highest fines content F_{c} of 10.0% in Fig. 3.5d, there were no clear peak point triggering the instability state and the PTS started to appear since the addition of C500 largely inhibited the contractancy and boosted the dilatancy, as previously mentioned in the above section.

Provided that the unification of the instability line can roughly be achieved for the test specimens under different consolidation stresses (100 kPa versus 400 kPa), the corresponding friction angles at the instability state listed in Table 3.2 were thereafter measured based on the test results under 400 kPa (see Fig. 3.6a) for better accuracy. Regarding F_{c} = 10.0%, the friction angle at PTS rather than that at the instability

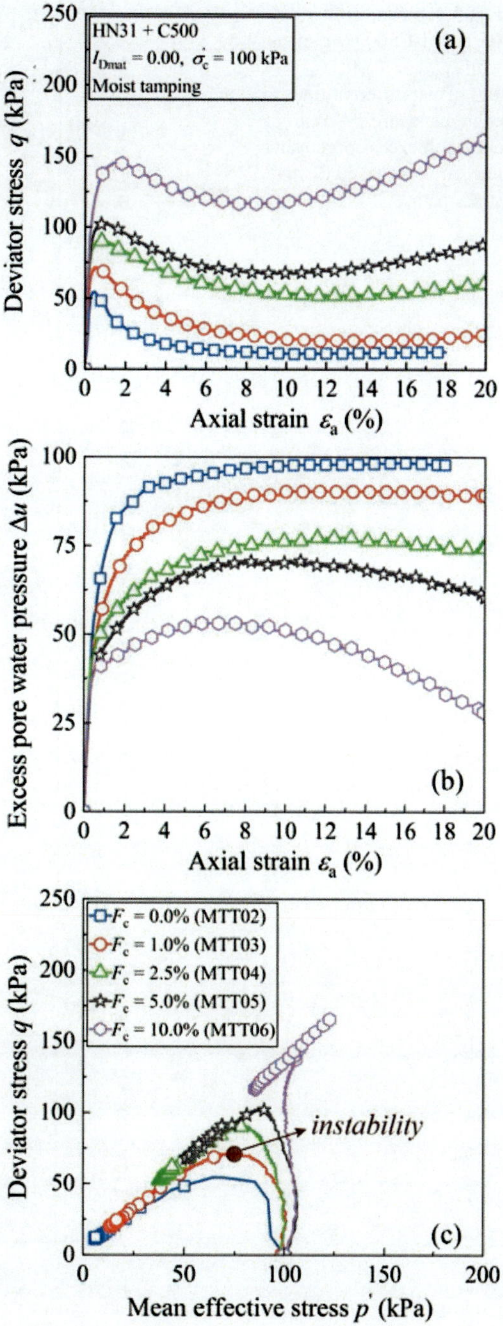

Fig. 3.3 Results obtained for HN31-C500 specimens with $I_{Dmat} = 0.00$, $\sigma'_c = 100$ kPa at different fines contents: **a** stress–strain curves; **b** excess pore water pressure curves; **c** effective stress paths

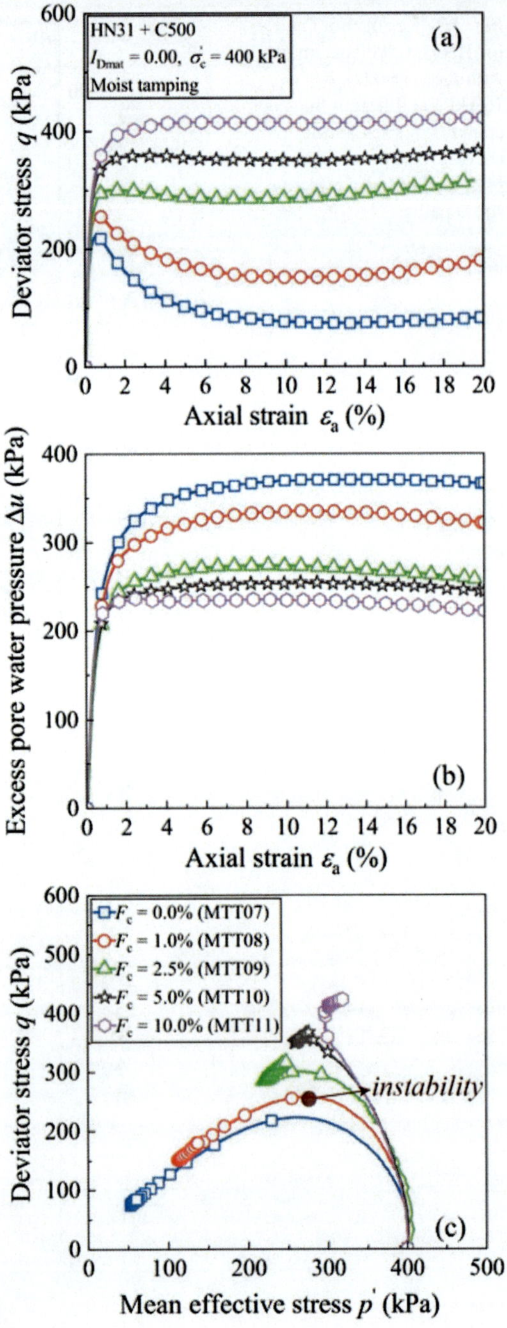

Fig. 3.4 Results obtained for HN31-C500 specimens with $I_{Dmat} = 0.00$, $\sigma'_c = 400$ kPa at different fines contents: **a** stress–strain curves; **b** excess pore water pressure curves; **c** effective stress paths

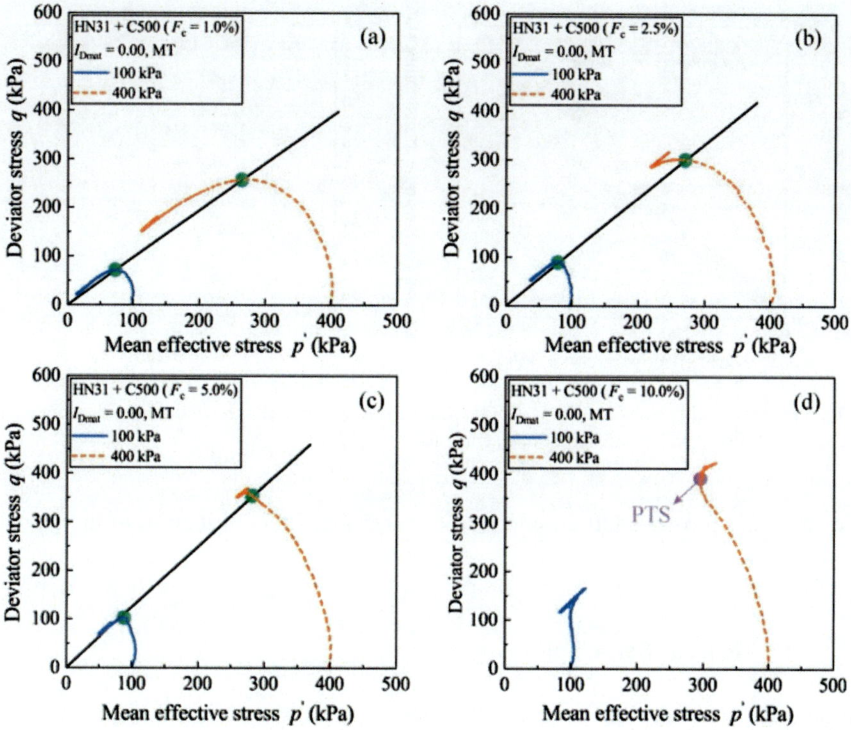

Fig. 3.5 Instability lines for HN31-C500 specimens with: **a** $F_c = 1.0\%$; **b** $F_c = 2.5\%$; **c** $F_c = 5.0\%$; **d** PTS line with $F_c = 10.0\%$

state was used to gain a rough idea of the upper boundary in Fig. 3.6b. For this reason, the concept for $F_c = 10.0\%$ was distinctively plotted by a discontinued line in Fig. 3.6a.

In Fig. 3.6b, it can be seen that $\phi^+{}_{inst}$ steadily increased and got closer to the PTS as F_c increased. This reconfirms that the addition of C500 silica continued to provide a positive contribution to the liquefaction resistance for the test specimens reconstituted with the moist tamping method in such a loose state. In addition, an exponential trend can be found in the graph. The initial addition of this type of fines was capable of efficiently and rapidly mitigating the liquefaction potential. In

Table 3.2 Friction angles mobilized at instability state for HN31-C500 specimens

HN31 + C500 ($I_{Dmat} = 0.00, \sigma'_c = 100$ kPa, Moist tamping)					
F_c (%)	$\phi^+{}_{inst}$ (°)	F_c (%)	$\phi^+{}_{inst}$ (°)	F_c (%)	$\phi^+{}_{PTS}$ (°)
0.0	20.2	2.5	27.9	10.0	32.9
1.0	23.6	5.0	28.8		

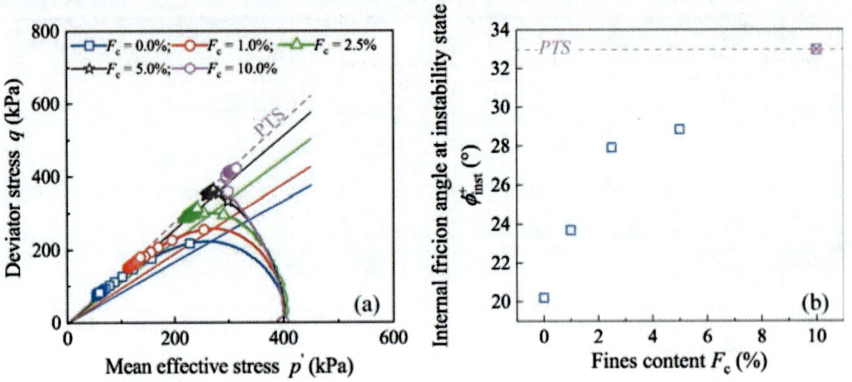

Fig. 3.6 **a** Instability lines for HN31-C500 specimens under 400 kPa; **b** friction angles mobilized at the instability state with different fines contents

contrast, the further addition provided a less beneficial contribution to the global force chain.

3.3.1.2 Medium-Dense State

Figure 3.7 shows the results obtained for HN31-C500 specimens ($I_{\mathrm{Dmat}} = 0.50$, $\sigma'_c = 100$ kPa) with different fines contents. As for stress–strain curves in Fig. 3.7a, the deviator stress q remarkably increased as the fines content F_c increased, revealing that the gradual addition of fine particles enhanced the global shear resistance. As for excess pore water pressure Δu curves, two phenomena could be identified: (i) before reaching the PTS ($\varepsilon_a < 0.5\%$), all five curves were almost overlapping with each other, revealing an insignificant effect of fine particles during this stage; and (ii) after passing through the PTS, the increase in fines content F_c favoured a dilatant behaviour, making Δu dissipated much more quickly. As for the effective stress paths depicted in Fig. 3.7c, all stress paths achieved the transition state from contractancy to dilatancy at the same single point. According to these observations, it can be deduced that the addition of C500 silica equally provided a quite positive contribution to these specimens in the medium-dense state, very similar to the conclusion drawn in the loose state of $I_{\mathrm{Dmat}} = 0.00$. Figure 3.8 presents the experimental results of HN31-C500 specimens with different fines contents under a consolidation stress equal to 400 kPa, allowing the same conclusions to be drawn.

Figure 3.9 presents a summary of all effective stress paths of HN31-C500 specimens in the medium-dense state of $I_{\mathrm{Dmat}} = 0.50$ under two consolidation stresses of 100 kPa and 400 kPa, respectively. In the graph, the attainments of PTS (outlined by the solid green circles) of all specimens can be unified by a single straight line. A slope of about 1.27 was suggested, pretty close to the value previously reported by Benahmed (2001). As is known to all, the PTS for a given granular material is commonly supposed to be intrinsic since its slope value is independent of the imposed

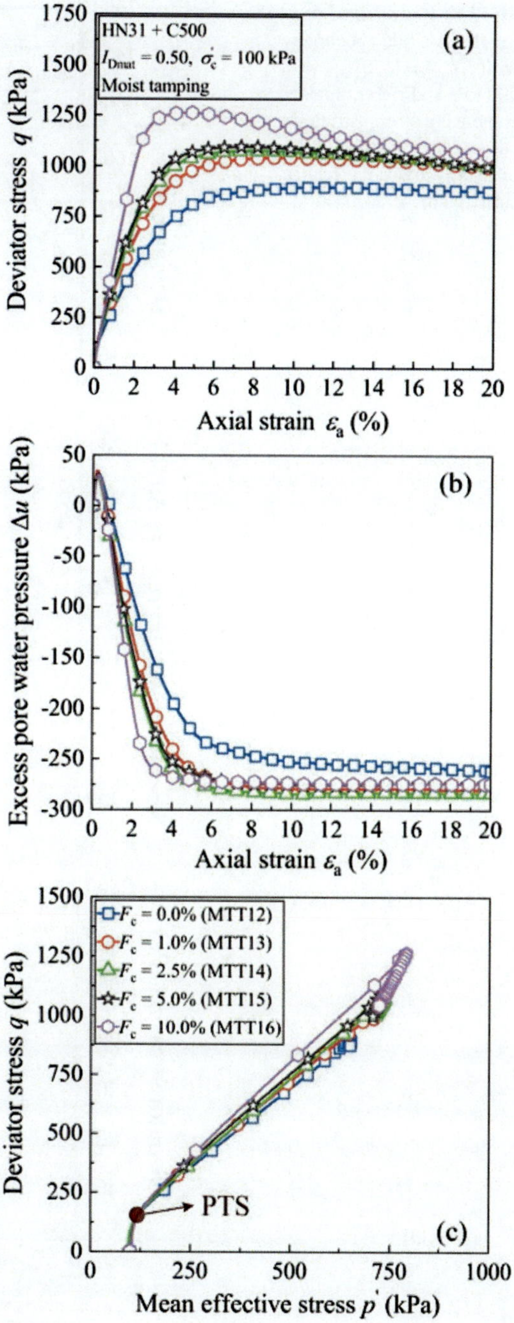

Fig. 3.7 Results obtained for HN31-C500 specimens with $I_{Dmat} = 0.50$, $\sigma'_c = 100$ kPa at different fines contents: **a** stress–strain curves; **b** excess pore water pressure curves; **c** effective stress paths

Fig. 3.8 Results obtained for HN31-C500 specimens with $I_{Dmat} = 0.50$, $\sigma'_c =$ 400 kPa at different fines contents: **a** stress–strain curves; **b** excess pore water pressure curves; **c** effective stress paths

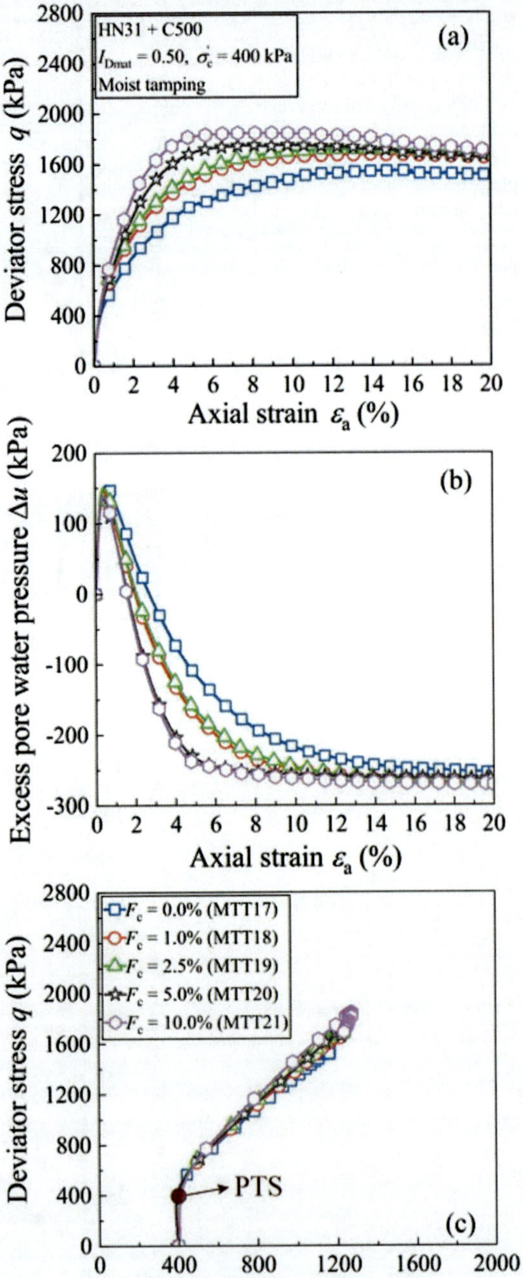

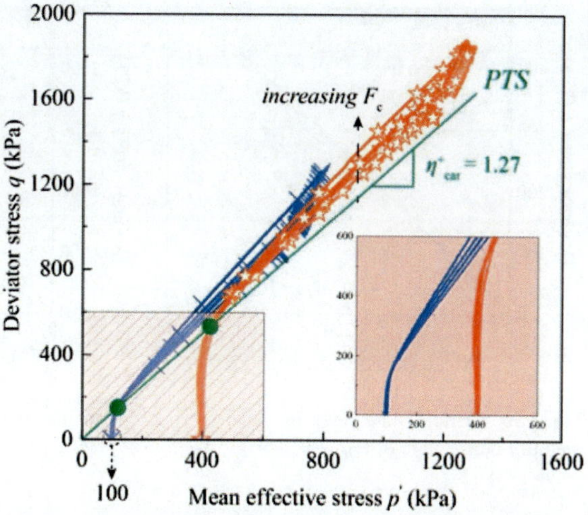

Fig. 3.9 PTS line for HN31-C500 specimens (moist tamping) in the medium-dense state of I_{Dmat} = 0.50 with different fines contents

loadings, density indexes and consolidation stresses. Besides these parameters and conditions, the finding in Fig. 3.9 reveals that the PTS is, furthermore, irrespective of the fines content F_c for the composite sandy specimen involving a prevailing sand matrix with the presence of small amounts of fine particles. This phenomenon can be logically explained by the fact that the PTS appeared in the initial phase (ε_a < 0.5%) in which the fine particles were almost inactive and only the main sand matrix resisted the external loading. That is the reason why the stress–strain curves and excess pore water pressure curves of HN31-C500 specimens almost coincided with those of clean sand specimens during the initial loading stage in Fig. 3.7 and Fig. 3.8, respectively.

3.3.1.3 Critical State Line

The critical state line is a 3D curve in the (q, p', e) space that can be projected in either the q-p' or e-logp' space. Figure 3.10 regroups the critical state lines of loose and medium-dense specimens under different consolidation stresses. Note that, for loose specimens, the critical state points were taken at the lowest value of the deviator stress before the dilatant behaviour took place. It can be noticed from Fig. 3.10 that the critical state point in the q-p' space increased with the increment in fines content for both density indexes. Furthermore, a good unification can be achieved. Some minor discrepancies between the medium-dense specimens can be ascribed to the fact that these specimens did not truly reach the critical state owing to the enhanced dilatant response with the addition of C500 silica.

The critical state lines in the e-logp' space with respect to global void ratio e are displayed in Fig. 3.11a. In the graph, a trend line in exponential form is used and its effectiveness has been proved by Zhu and Cheng (2020). It can be seen that each mixture constructs its own critical state line in this space; thereby, the unification

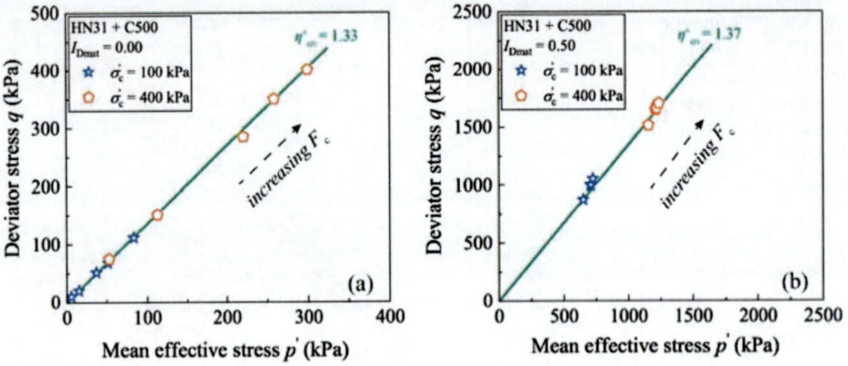

Fig. 3.10 Critical state lines in the q-p' space: **a** in the loose state of $I_{\mathrm{Dmat}} = 0.00$; **b** in the medium-dense state of $I_{\mathrm{Dmat}} = 0.50$

becomes no longer accomplished. In addition, a clockwise rotation is observed. This phenomenon can be explained by the following two independent reasons. First, with the addition of C500 silica, the dilatant behaviour leading to a greater value of mean effective stress at the final state was enhanced. Therefore, a moving trend toward the right can be expected. Second, by using the global void ratio as the state parameter (including both coarse sand grains and fine particles), the specimens were algebraically densified at a higher fines content. Therefore, a downward movement can be imagined.

The critical state line in the e-logp' space with respect to the equivalent void ratio e^{eq} (in Table 3.3) is shown in Fig. 3.11b. The evolution of $e_{\max}/e_{\min}$ with fines content was shown in Zhu et al. (2021a, b) and the Eq. 1.5 presented in *Chap.* 1 is here used to determine the value of b which stands for the active fraction of fine particles that may effectively participate in global force chain.

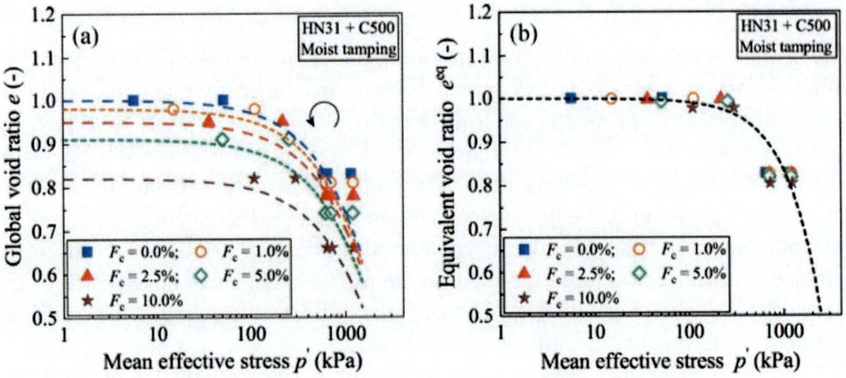

Fig. 3.11 Critical state lines in the e-logp' space in terms of: **a** global void ratio; **b** equivalent void ratio

Table 3.3 Equivalent void ratios for HN31-C500 specimens (moist tamping)

Reference	I_{Dmat}	F_c [(1)] (%)	f_c [(2)] (%)	e_{mat}	e	e_s [(3)]	$e_{\max}$ [(4)]	$e_{\min}$ [(5)]	b [(6)]	e^{eq} [(7)]
MTT01	0.00	0.00	0.00	1.00	1.00	1.00	1.00	0.66	0.06	1.000
MTT03		1.00	0.99	1.00	0.98	1.00	1.01	0.65	0.06	0.999
MTT04		2.50	2.44	1.00	0.95	1.00	1.03	0.63	0.07	0.997
MTT05		5.00	4.76	1.00	0.90	1.00	1.06	0.6	0.08	0.992
MTT06		10.00	9.09	1.00	0.82	1.00	1.09	0.51	0.11	0.977
MTT07	0.00	0.00	0.00	1.00	1.00	1.00	1.00	0.66	0.06	1.000
MTT08		1.00	0.99	1.00	0.98	1.00	1.01	0.65	0.06	0.999
MTT09		2.50	2.44	1.00	0.95	1.00	1.03	0.63	0.07	0.997
MTT10		5.00	4.76	1.00	0.90	1.00	1.06	0.6	0.08	0.992
MTT11		10.00	9.09	1.00	0.82	1.00	1.09	0.51	0.11	0.977
MTT12	0.50	0.00	0.00	0.83	0.83	0.83	1.00	0.66	0.07	0.828
MTT13		1.00	0.99	0.83	0.81	0.83	1.01	0.65	0.07	0.827
MTT14		2.50	2.44	0.83	0.78	0.83	1.03	0.63	0.08	0.824
MTT15		5.00	4.76	0.83	0.74	0.83	1.06	0.6	0.10	0.819
MTT16		10.00	9.09	0.83	0.66	0.83	1.09	0.51	0.13	0.804
MTT17	0.50	0.00	0.00	0.83	0.83	0.83	1.00	0.66	0.07	0.828
MTT18		1.00	0.99	0.83	0.81	0.83	1.01	0.65	0.07	0.827
MTT19		2.50	2.44	0.83	0.78	0.83	1.03	0.63	0.08	0.824
MTT20		5.00	4.76	0.83	0.74	0.83	1.06	0.6	0.10	0.819
MTT21		10.00	9.09	0.83	0.66	0.83	1.09	0.51	0.13	0.804

Note (1) $F_c = m_{\mathrm{fines}}/m_{\mathrm{sand}}$, (2) $f_c = m_{\mathrm{fines}}/(m_{\mathrm{sand}} + m_{\mathrm{fines}})$, (3) e_s: skeleton void ratio, (4, 5) according to Zhu et al. (2021a, b), (6) b according to Gobbi et al. (2022a, b), (7) e^{eq}: equivalent void ratio

It can be seen from Fig. 3.11b that all experimental data roughly converge on a single exponential curve, irrespective of the value of the fines content F_c. This is consistent with the experimental results presented in *Chap.* 1 that while assessing the critical state line of sandy specimens containing fines particles, a unified critical state can be satisfactorily achieved with the use of equivalent void ratio.

3.3.2 *Influence of C500 Silica Using the Dry Tamping Method*

Owing to the fact that C500 silica is non-plastic, this type of fine particles is compatible with two reconstitution methods (dry tamping and moist tamping), providing the possibility to focus on the influence of initial soil fabric created by different methods. Figure 3.12 shows the experimental results obtained for HN31-C500 specimens ($I_{Dmat} = 0.00$, $\sigma'_c = 100$ kPa and dry tamping method) with different fines contents. As for the stress–strain curves in Fig. 3.12a, it can be seen that the deviator stress q steadily increased as the axial loading was applied and did not exhibit any downward trend, signifying a low potential of flow liquefaction. And the increase in fines content F_c remarkably weakened the shear resistance. As for the excess pore water pressure Δu curves in Fig. 3.12b, before the attainment of PTS, all curves seemed to be the same. After passing through this state, the increase in F_c suppressed the soil dilatancy, making Δu dissipated much less. As for the effective stress paths displayed in Fig. 3.12a, the achievement of the PTS appeared to be irrespective of the F_c and all stress paths seemed to follow the same rupture line.

The above phenomena suggest that for HN31-C500 specimens reconstituted with the dry tamping method, the addition of C500 silica provides a negative contribution to the global force chain. This is opposite to the observation for HN31-C500 specimens fabricated with the moist tamping method in which the gradual addition of C500 silica enhanced the undrained shear resistance.

For simplicity, the undrained shear strength S_u, defined as half the maximum deviator stress q_{max} during shearing, is used as a comparison basis to assess the influence of fines content F_c. The relationship between S_u and F_c is shown in Fig. 3.13a in which a linear decreasing tendency can be found, revealing again a negative contribution of fine particles to the global force chain.

While assessing sandy specimens containing fine particles, the key hypothesis behind the concept of equivalent void ratio e^{eq} is its ability to unify the mechanical response in order that two specimens should behave very similarly if both are prepared at the same value of e^{eq}, which is furthermore irrespective of F_c. In like manner, this concept assumes that if two specimens are prepared at two different values of e^{eq}, the specimen owing the smaller e^{eq} (densification) should be more resistant than the other, and vice versa. To examine whether this key hypothesis can hold true, the values of e^{eq} for HN31-C500 specimens reconstituted with the dry tamping method (Table 3.4) were also determined using the method presented in the above section.

In Fig. 3.13b, it can be seen that with the increment in F_c (from 5.0 to 15.0%), the corresponding e^{eq} slightly decreases standing for a densification, whilst S_u significantly decreases standing for a loosening. This logical inconsistency suggests that the equivalent void ratio is not a rational state variable for characterizing sand-fines specimens reconstituted with the dry tamping method. The reason to explain this anomaly might be the fact that fine particles in this concept are supposed to always provide a positive contribution to the global force chain with a variable b varying between 0 and 1. However, the applied reconstitution method can sometimes reverse the influence of fine particles, which is not fully considered in the current framework so far.

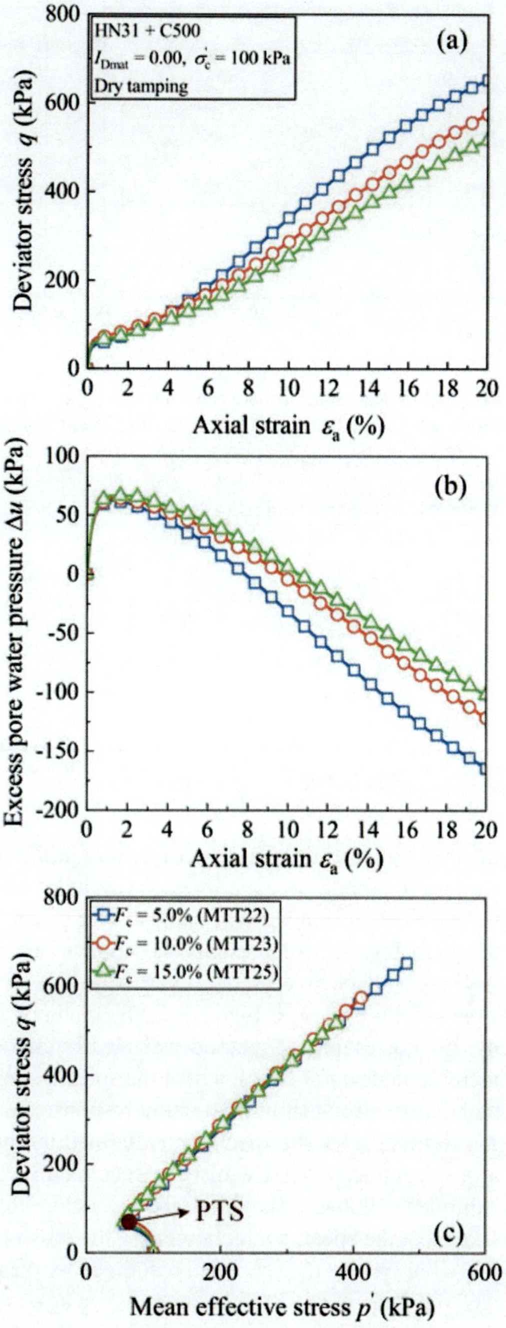

Fig. 3.12 Results obtained for HN31-C500 specimens with $I_{Dmat} = 0.00$, σ'_c = 100 kPa at different fines contents: **a** stress–strain curves; **b** excess pore water pressure curves; **c** effective stress paths

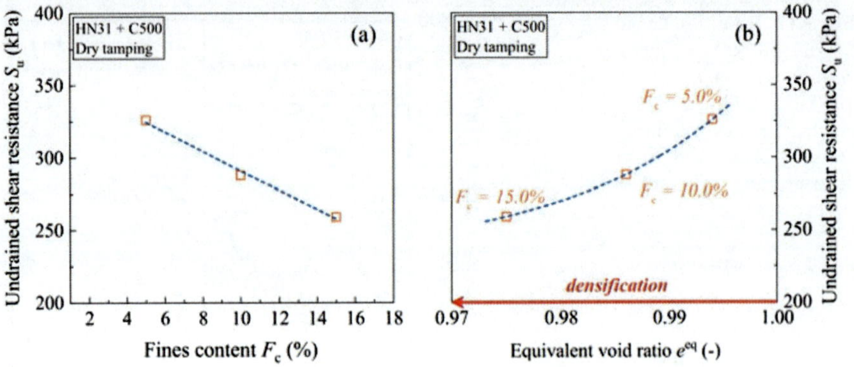

Fig. 3.13 **a** Relationship between undrained shear resistance and fines content; **b** relationship between undrained shear resistance and equivalent void ratio

Table 3.4 Equivalent void ratios for HN31-C500 specimens (dry tamping)

Reference	I_{Dmat}	F_c (%)	f_c (%)	e_{mat}	e	e_s	e_{max}	e_{min}	b	e^{eq}
MTT22	0.00	5.00	4.76	1.00	0.90	1.00	1.00	0.66	0.06	0.994
MTT23		10.00	9.09	1.00	0.82	1.00	1.01	0.65	0.07	0.986
MTT25		15.00	13.04	1.00	0.74	1.00	1.03	0.63	0.09	0.975

3.3.3 Influence of Reconstitution Method for Sand-C500 Mixtures

In order to clearly emphasize the effect of initial soil fabric associated with the reconstitution method, the experimental results of HN31-C500 specimens reconstituted by two methods but with the same testing parameters (I_{Dmat}, σ'_c and F_c) are plotted together. Figure 3.14 presents the comparison of the experimental results of two HN31-C500 specimens ($I_{Dmat} = 0.00$, $\sigma'_c = 100$ kPa and $F_c = 5.0\%$). As for the stress–strain curves in Fig. 3.14a, the deviator stress q of the specimen reclaimed by the moist tamping method was clearly characterised by a drop after the initial increase in deviator stress, whilst the specimen reclaimed by the dry tamping exhibited a quite strong strain-hardening response. As for the excess pore water pressure Δu in Fig. 3.14b, the specimen reclaimed by the moist tamping method displayed a pure contractiveness, whilst the specimen reclaimed by the dry tamping method exhibited a dilatant behaviour. The corresponding effective stress paths in Fig. 3.14c show that the specimen reclaimed by the moist tamping was governed by the instability line, whilst the specimen reclaimed by the dry tamping method followed quite well the concept of PTS. Therefore, it can be concluded that the soil fabric shaped by the moist tamping method is much more liquefiable than that with the dry tamping method. The same conclusion can also be drawn for a similar comparison at $F_c = 10.0\%$ in Fig. 3.15.

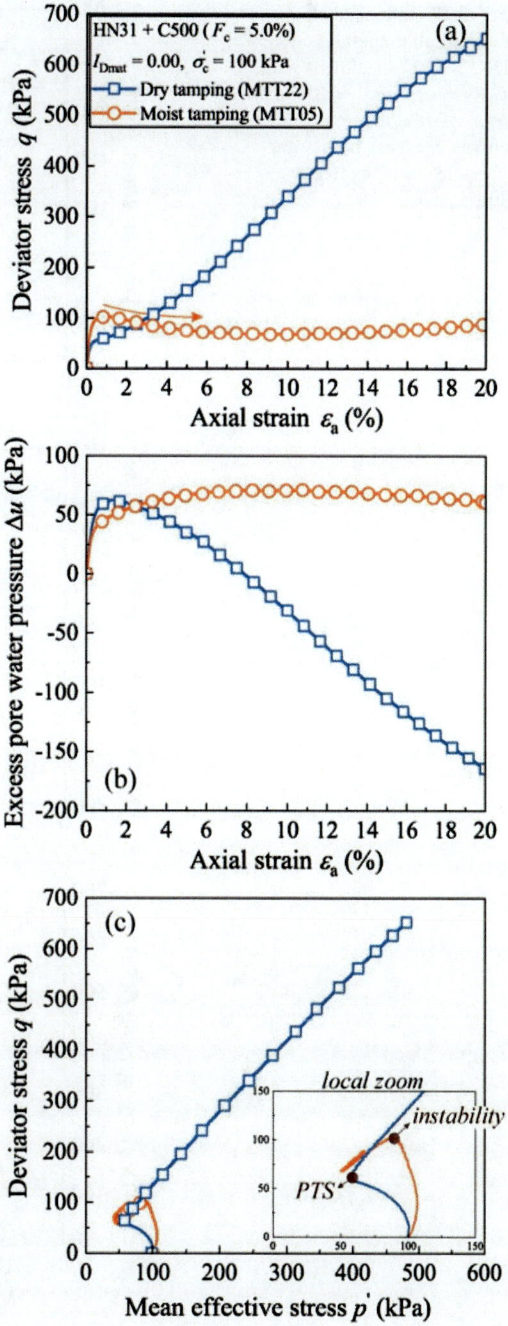

Fig. 3.14 Influence of reconstitution method on HN31-C500 specimens with $I_{Dmat} = 0.00$, $\sigma'_c = 100$ kPa, $F_c = 5.0\%$: **a** stress–strain curves; **b** excess pore water pressure curve; **c** effective stress paths

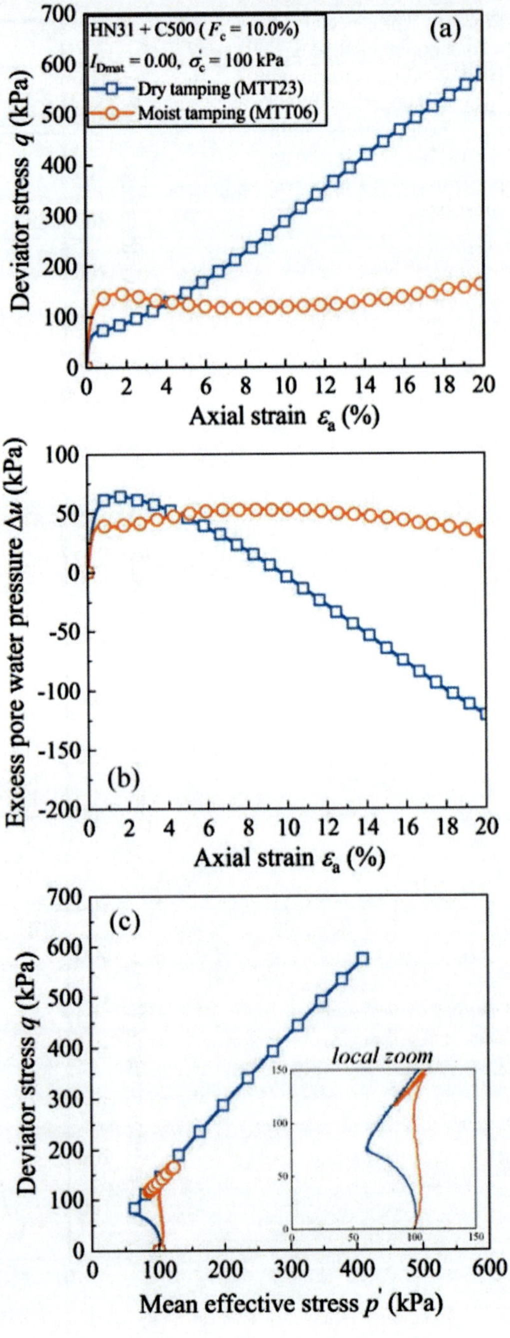

Fig. 3.15 Influence of reconstitution method on HN31-C500 specimens with $I_{\text{Dmat}} = 0.00$, $\sigma'_c = 100$ kPa, $F_c = 10.0\%$: **a** stress–strain curves; **b** excess pore water pressure curve; **c** effective stress paths

The above misleading issues can only be ascribed to the initial soil fabric since all other testing parameters were kept constant while reconstituting the specimens. For this purpose, microstructural observations were performed and the objective was to directly visualize the initial soil arrangement formed with two different reconstitution methods. Owing to the fact that the colours of two tested materials are highly similar due to the similar mineral composition ($S_iO_2 > 99\%$), it is challenging to distinguish the coarse sand grains from fine particles, especially on a relatively small scale. For the sake of a proper distinction, the coarse HN31 sand was, in particular, dyed blue using a commercially available colourant (BASACID Blue 762 liquid supplied by BASF), as shown in Fig. 3.16. Afterwards, the fresh dyed HN31 sand was carefully placed in a huge oven for more than an hour at a constant temperature of 105 °C to maintain its colour further. Finally, the grain size distribution curve of coloured HN31 sand was roughly re-measured by classic dry-sieving. The result showed that neither particle size nor sand grain aggregation occurred during the dyeing procedure described above.

In order to represent as much as possible the initial fabric of sand-fines mixtures corresponding to one of the specimens installed into the triaxial apparatus before the saturation phase, the composite soil ($I_{\mathrm{Dmat}} = 0.00$) was deposited in a see-through cylindrical cup of 70 mm in diameter and 14 mm (see Fig. 3.17a) in height by dry and moist tamping method, respectively. Microstructural observations were thereafter performed using a high-resolution optical microscope (see Fig. 3.17b). Due to the intrinsic constraints of two reconstitution methods, microstructural observations were only performed on two fines contents F_c of 5.0% and 10.0%, respectively.

Figure 3.18a, b present the microstructural observations on HN31-C500 specimens stacked up in dry state at F_c of 5.0 and 10.0% in which the black/white/blue regions graphically represent the inter-sand void, the stacked up C500 silt and the coloured HN31, respectively. It can be observed from two images that the added fine

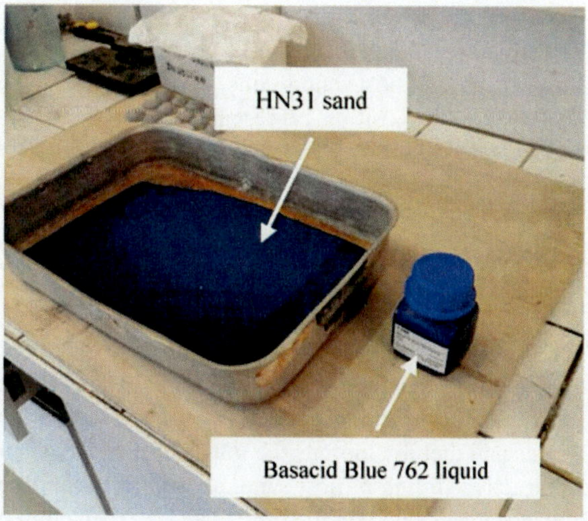

Fig. 3.16 HN31 sand coloured with basacid blue 762 liquid

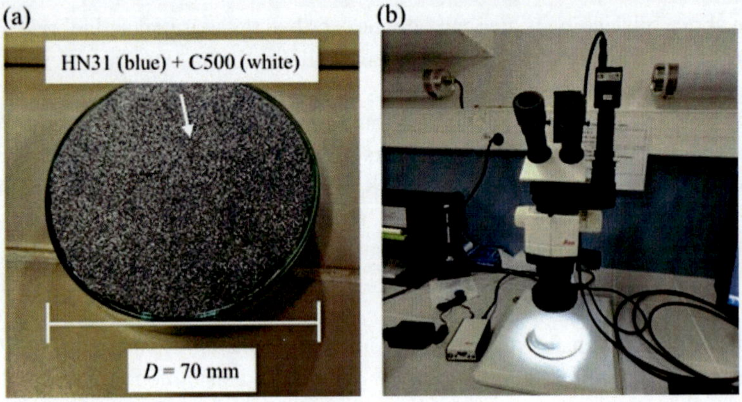

Fig. 3.17 **a** Sand-fines mixture with coloured HN31 sand; **b** high-resolution optical microscope

particles not only filled the inter-sand voids but also located ***nearby the sand-sand contact points by separating them***. Similarly, Fig. 3.19a, b show the microstructural observations for HN31-C500 specimens prepared by the moist tamping method at F_c of 5.0% and 10.0%, respectively. It appears from two figures that: (i) the fines were highly bulked together, and this phenomenon became more pronounced as the fines content increased; (ii) the coarse HN31 sand was remarkably covered by the bulked fines with very few fine particles falling in the inter-sand voids, which could be attributed to the capillary force while employing the moist tamping method; (iii) the coarse sand with the fine particles were generally still in contact with one another, and the initial global fabric of this structure was irregular and primarily governed by the aggregated coarse sand forming large metastable macropores in such a loose state. In an experimental study of microstructure of FSand N34-K clay mixture (equally reconstituted in the presence of water with water content $w_L = 1.5$), scanning electron microscopy (SEM) has been recently applied by Yin et al. (2021) to check the soil structure. In Fig. 3.20, the K clay highly stuck together and formed a continuous thin film around host sand grains, very similar to the present work.

For a clear exposition, representations of different structures of HN31-C500 specimen created by the dry and moist tamping method are shown in Figs. 3.21 and 3.22 where large hollow and small solid circles represent the coarse HN31 sand and C500 silica, respectively. In the case of the dry tamping method, the fine particles not only (i) fill the inter-sand voids, (ii) but also locate nearby the sand-sand contact points by separating them since the mean diameter of C500 silica is far smaller than that of the coarse HN31 sand. In the first case (involved in the inter-sand voids), the fine particles make no actual mechanical contribution to the soil response and could be viewed as inactive since they are totally free to shift. By contrast, in the second case (nearby the contact points), the presence of C500 silica disperses the coarse grains and thereby loosens the initial host matrix. This mechanism can facilitate the contractive sliding of host sand, as well as inhibit the dilatant response. Moreover, this negative or weakening effect can be expected to become much more significant

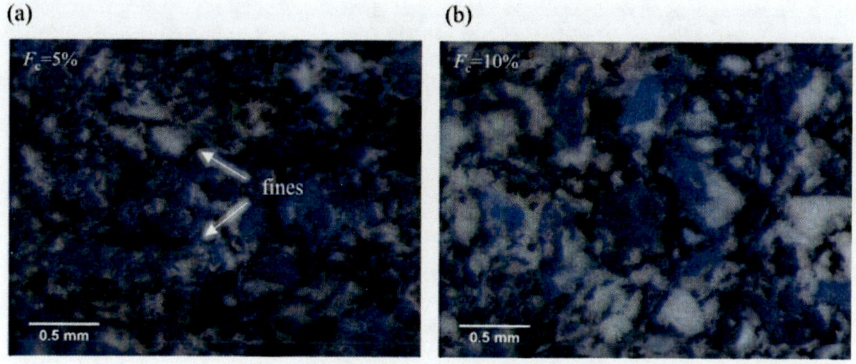

Fig. 3.18 Microstructural observations of sand-fines mixtures in very loose state reconstituted with the dry tamping method: **a** $F_c = 5\%$; **b** $F_c = 10\%$. (For interpretation of the references to colour in this figure, the reader is referred to the web version of this thesis)

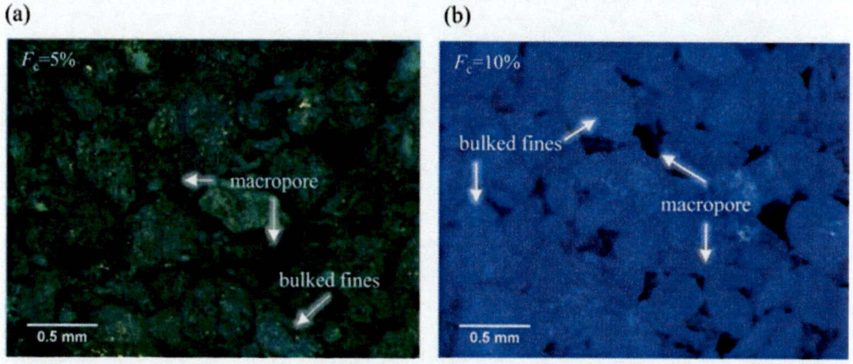

Fig. 3.19 Microstructural observations of sand-fines mixtures in very loose state reconstituted with the moist tamping method: **a** $F_c = 5\%$; **b** $F_c = 10\%$. (For interpretation of the references to colour in this figure, the reader is referred to the web version of this thesis)

at higher fines content. This can explain why the deviator stress decreased with the increase in fines content, as shown in Fig. 3.12.

In the case of the moist tamping method depicted in Fig. 3.22, the presence of macropores makes the initial soil fabric exhibit an extremely high collapsibility potential and these macropores tend to decrease in volume upon shearing, which results in a rapid generation of excess pore water pressure under undrained condition. As a result, the overall test specimen becomes more sensible to static liquefaction. This can logically explain why composite specimens reconstituted with the moist tamping method exhibited a more or less strain-softening or liquefaction response, as shown in Figs. 3.14 and 3.15. However, the presence of C500 silica cannot truly disperse the internal contact between adjacent sand grains since they are closely attached to the coarse sand grains. As long as the host sand matrix yielding the

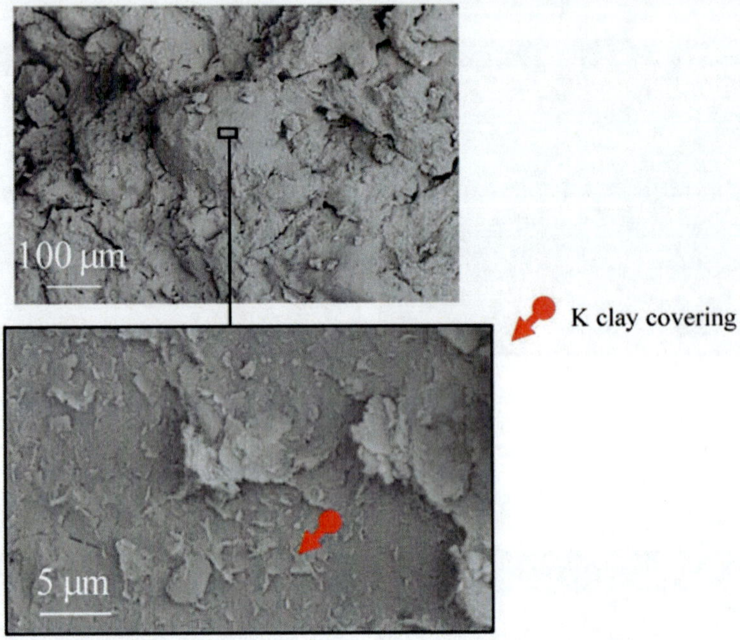

Fig. 3.20 SEM images in secondary electron imaging mode of specimens (composed of 50 wt.% K clay and 50 wt.% F sand NE34 with initial water content $w_L = 1.5$) (after Yin et al. 2021)

Fig. 3.21 Schematic of hypothesized composite soil fabric reconstituted with the dry tamping method

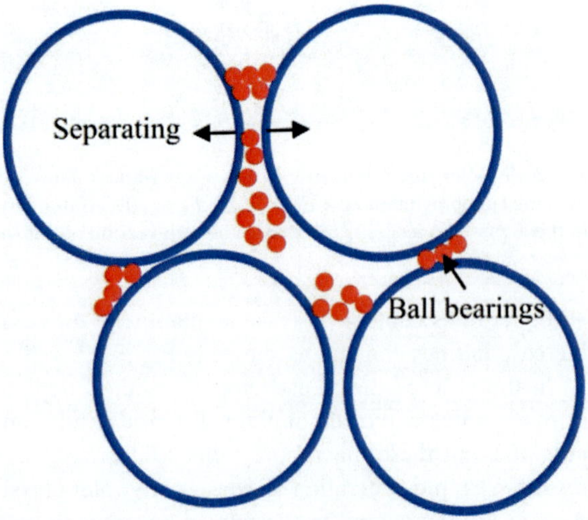

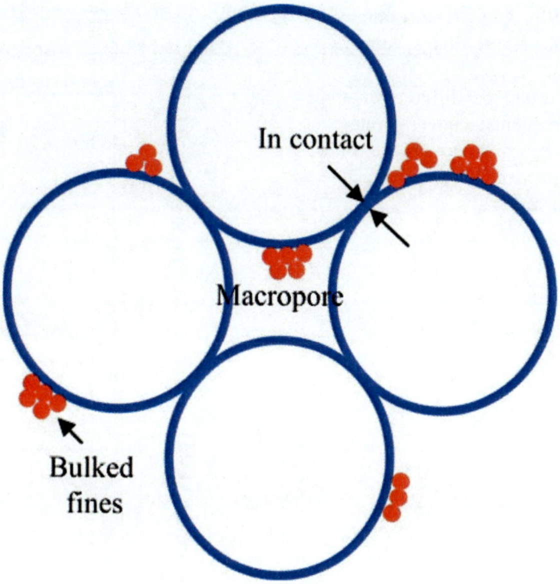

Fig. 3.22 Schematic of hypothesized composite soil fabric reconstituted with the moist tamping method

global force chain is kept almost unchanged with the addition of fine particles, the basic soil mechanics usually suggests an enhanced shear resistance while densifying the test specimen through introducing fine particles. This can logically explain why the mechanical role of C500 silica is interestingly beneficial and provides a very positive contribution to the global shear resistance for this granular configuration, which is totally different from that observed in the case of the dry tamping method.

3.3.4 Influence of Plastic Fine Particles Using the Dry Tamping Method

Figures 3.23 and 3.24 present the experimental results obtained for HN31-Illite and HN31-Speswhite specimens ($I_{Dmat} = 0.00$, $\sigma'_c = 100$ kPa) with different fines contents, respectively. Regarding the influence of fine particles in the case of the ***dry tamping method***, the increment in fines content F_c also provided a negative contribution to the shear resistance due to the following observations. As for the stress–strain curves in Figs. 3.23a and 3.24a, the deviator stress q obviously decreased with the increase in fines content F_c. Similarly, the excess pore water pressure Δu increased as F_c increased in Figs. 3.23b and 3.24b, revealing an enhanced contractive response with the addition of these two plastic fines.

Here, it becomes attractive to emphasize that the significant increase in plasticity index I_p (from NP of C500 to $I_p = 34$ of Illite and to $I_p = 30$ of Speswhite) does not alter the conceptual models detailed in the above section. As for a granular structure

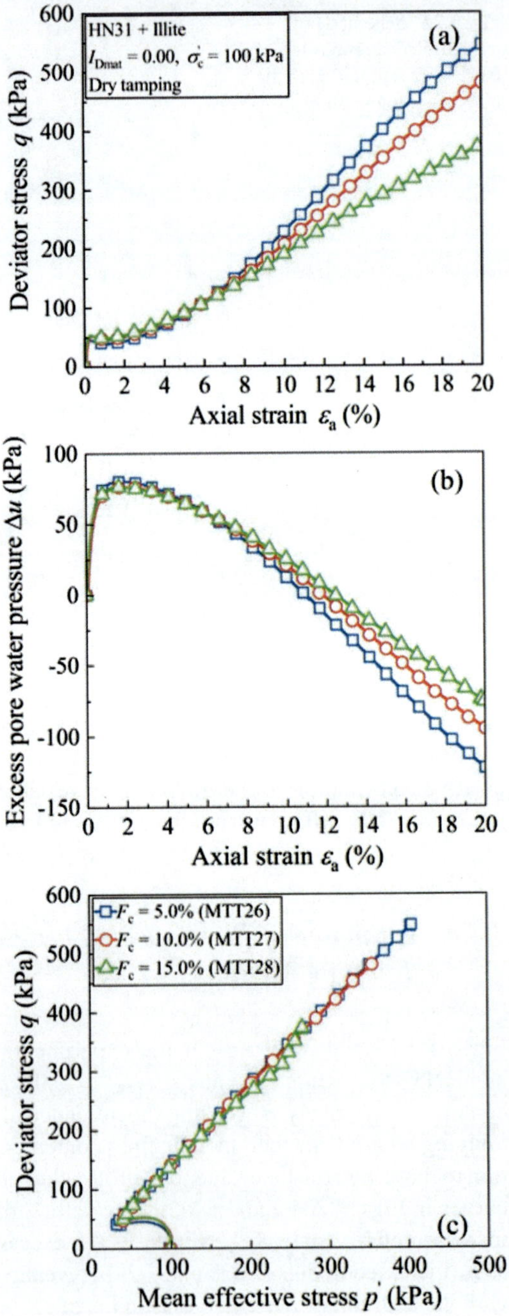

Fig. 3.23 Results obtained for HN31-Illite specimens with $I_{Dmat} = 0.00$, $\sigma'_c = 100$ kPa at different fines contents: **a** stress–strain curves; **b** excess pore water pressure curves; **c** effective stress paths

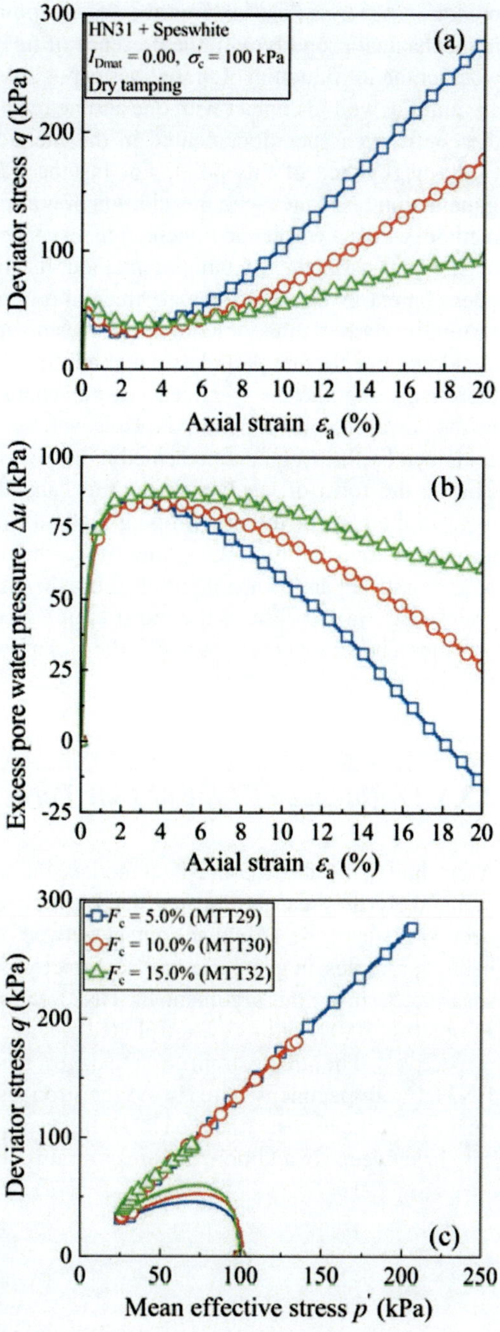

Fig. 3.24 Results obtained for HN31-speswhite specimens with $I_{\mathrm{Dmat}} = 0.00$, $\sigma'_c = 100$ kPa at different fines contents: **a** stress–strain curves; **b** excess pore water pressure curves; **c** effective stress paths

predominated by a host sand matrix, the key point to assess its subsequent variation in mechanical property with the presence of fine particles should be the question of whether the introduction of these fines is possible to disturb the host sand grains that are initially well in contact with one another, yielding the global force chain. In fact, the confusing issues documented in the literature might be sometimes attributed to the negligence of this point. For instance, Jradi (2018) reported an enhanced liquefaction resistance with the addition of non-plastic fines using the moist tamping method, whilst a weakened liquefaction resistance was observed with the addition of plastic fines using the dry tamping method. It is of interest to note that the maximum fines content involved in the experimental programme was only 5%. Although it is straightforward to think of a growing cohesion against the static or flow liquefaction by adding plastic fine particles, a maximum value of only 5% remains too little to thoroughly alter the test specimen's overall characteristics, which is mainly governed by the prevailing host sand matrix. However, the above observations can be logically explained by the argument presented in this work. If small amounts of fine particles exist in the form of weakening the mechanical property of the host sand matrix (e.g., in the case of the dry tamping method), the addition of these fines provides a negative contribution to the global force chain. On the contrary, if small amounts of fine particles are incapable of breaking the effective contact between coarse sand grains (e.g., in the case of the moist tamping method), the addition of these fines provides a beneficial contribution to the global force chain, as suggested by the basic soil mechanics.

3.3.5 *Influence of Type of Fine Particles*

According to the above conceptual models, the smaller the fine particles become, the more likely they can be located nearby the sand-sand contact points and subsequently destroy the effective contact between coarse sand grains. This rationale behind promotes, in general, a more contractive sliding for the host sand matrix upon shearing. To prove this argument, the HN31-fines specimens at the same fines content F_c of 5.0%, 10.0% and 15.0% are plotted together in Fig. 3.25, Fig. 3.26 and Fig. 3.27, respectively. The stress–strain curves suggest that, at the same testing parameters, the HN31-C500 specimens were always the strongest, whilst the HN31-Speswhite specimens the lowest. And the HN31-Illite specimens could be considered to be the case in between . In addition, the stress–strain curves for HN31-Speswhite specimens were clearly characterized by a clear drop after the initial increase in deviator stress q, referring to a high contractiveness. By contrast, as the fine particles became significantly coarser, the stress–strain curves for HN31-C500 specimens steadily increased without any drop tendency, indicating a strong dilatancy. The similar conclusion can also be drawn from the excess pore water pressure curves and the corresponding effective stress paths, respectively.

In the published literature, a common viewpoint advises that the variation in undrained shear resistance of sand-fines mixtures can be expressed as a function of the plasticity index I_p of the added fines. For instance, the addition of plastic fines is adverse to the global force chain, whilst a very positive contribution shall be expected with the gradual addition of non-plastic fines (Jradi 2018; Jradi et al. 2021). Figure 3.25 can verify whether the above idea could hold true. While passing through non-plastic C500 to plastic Speswhite of $I_p = 30$, the increase in I_p indeed made the specimen exhibit a more contractive or liquefaction tendency. Following up this logic, the next increment in I_p (from $I_p = 30$ of Speswhite to $I_p = 34$ of Illite) should continue to amplify the liquefaction potential. Unfortunately, the HN31-Illite specimens were interestingly much stronger than the HN31-Speswhite specimens at all three F_c of 5.0% in Fig. 3.25, 10.0% in Fig. 3.26 and 15.0% in Fig. 3.27. Such an observation is contrary to the above key underlying assumption. The shortcoming of I_p is evident since this individual parameter only considers the property of fines themselves and cannot express the physical manner in which fine particles are deposited to a given predominant host sand matrix. As an alternative, the mean diameter ratio D_{50}/d_{50}, defined as the mean diameter of the coarse grain D_{50} divided by the mean diameter of fine particles d_{50}, is thereafter adopted to interpret the test results.

For simplicity, the undrained shear strength S_u is reused to assess the performance of D_{50}/d_{50}. Figure 3.28 presents the relationship between S_u and F_c for all tested HN31-fines specimens reconstituted with the dry tamping method. In the graph, a unique tendency about D_{50}/d_{50} can be found at all three fines contents: the greater D_{50}/d_{50} becomes, the smaller S_u of the HN31-fines specimen attains. By contrast, S_u initially decreases with the increment in I_p and subsequently rises again with the further incremental step in I_p. The advantageous point of D_{50}/d_{50} over other individual parameters (e.g., I_p) only representing fine particles is the consideration of the combined influence of both the coarse sand grains and the added fines. As fine particles get smaller leading to a greater value of D_{50}/d_{50}, they are more likely to be located nearby the sand-sand contact points by separating them. As a result, the host sand matrix yielding the global force chain is more likely to be damaged. This microscopic viewpoint can reasonably explain the unique tendency in regard to D_{50}/d_{50} in Fig. 3.28.

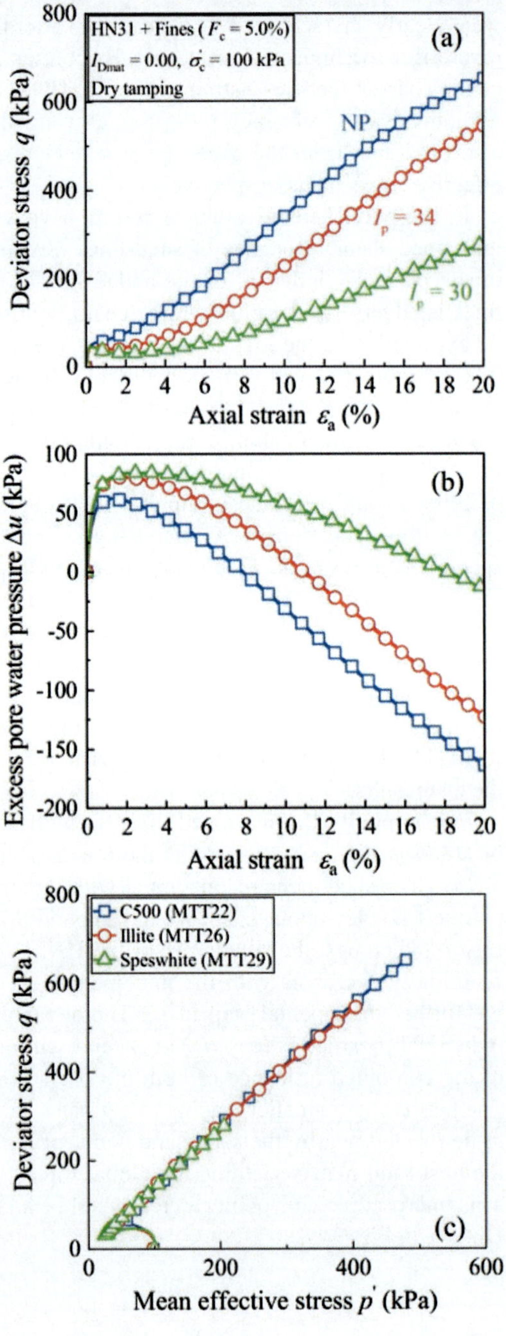

Fig. 3.25 Influence of mean diameter ratio on HN31-fines specimens with $I_{Dmat} = 0.00$, $\sigma'_c = 100$ kPa, $F_c = 5.0\%$: **a** stress–strain curves; **b** excess pore water pressure curve; **c** effective stress paths

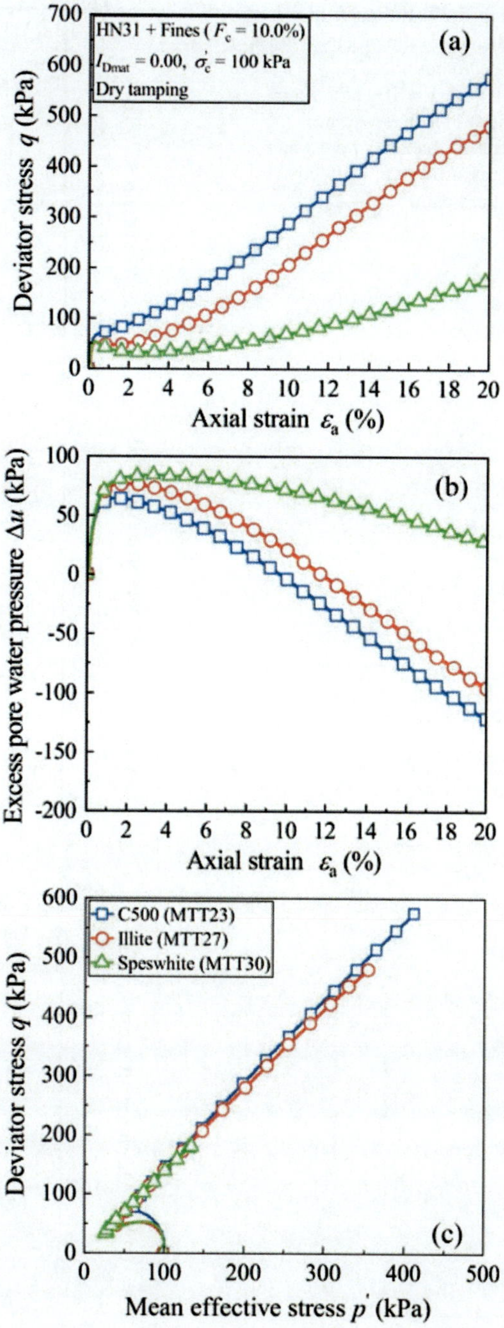

Fig. 3.26 Influence of mean diameter ratio on HN31-fines specimens with $I_{Dmat} = 0.00$, $\sigma'_c = 100$ kPa, $F_c = 10.0\%$: **a** stress–strain curves; **b** excess pore water pressure curve; **c** effective stress paths

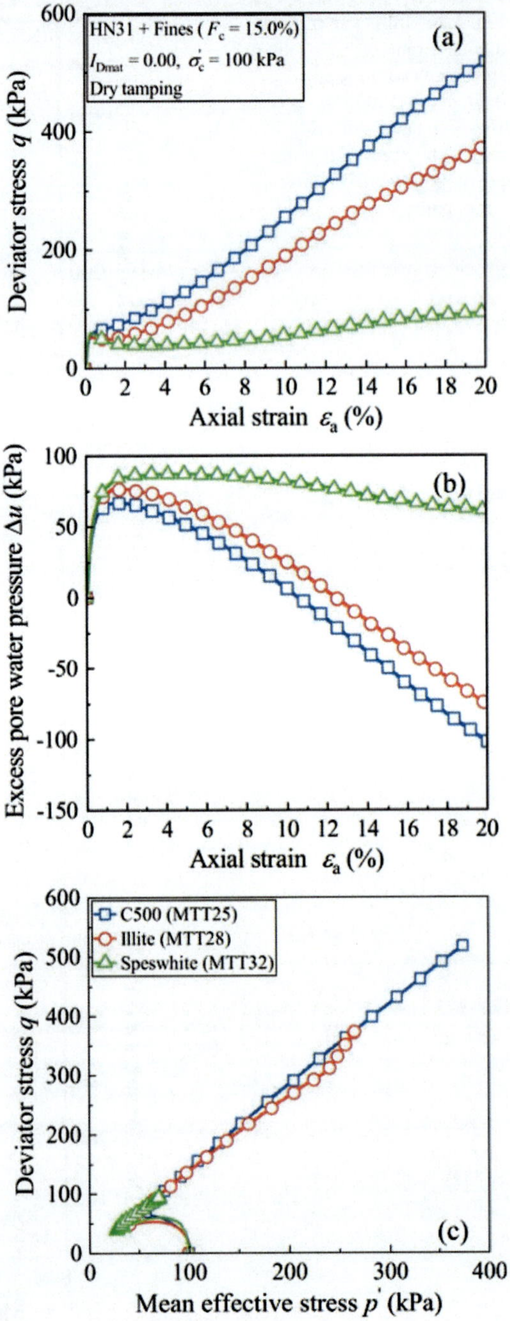

Fig. 3.27 Influence of mean diameter ratio on HN31-fines specimens with $I_{Dmat} = 0.00$, $\sigma'_c = 100$ kPa, $F_c = 15.0\%$: **a** stress–strain curves; **b** excess pore water pressure curve; **c** effective stress paths

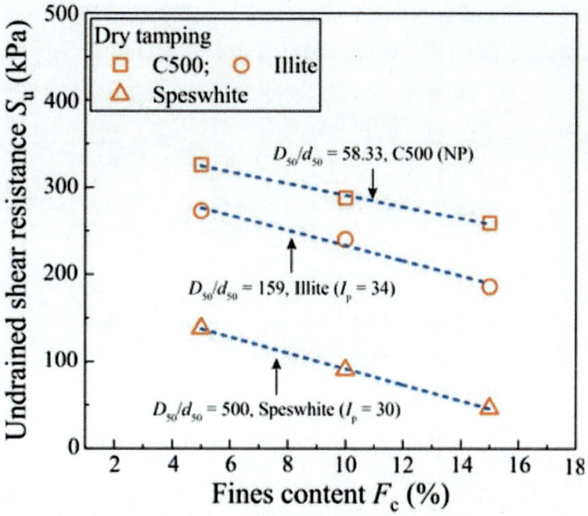

Fig. 3.28 Influence of mean diameter ratio D_{50}/d_{50} on undrained shear strength

References

Benahmed N (2001) Comportement mécanique d'un sable sous cisaillement monotone et cyclique: application aux phénomènes de liquéfaction et de mobilité cyclique. Ph.D Thesis Ecole Nationale Des Ponts et Chaussées

Gobbi S, Reiffsteck P, Lenti L, d'Avila MPS, Semblat JF (2022a) Liquefaction triggering in silty sands: effects of non-plastic fines and mixture-packing conditions. Acta Geotechnica 17(2):391–410. Springer Berlin Heidelberg. https://doi.org/10.1007/s11440-021-01262-1

Gobbi S, Santisi d'Avila MP, Lenti L, Semblat J-F, Reiffsteck P 2022b Effect of active plastic fine fraction on undrained behavior of binary granular mixtures. Int J Geomech 22(1). https://doi.org/10.1061/(asce)gm.1943-5622.0002242

Jradi L, El Dine BS, Dupla JC, Canou J (2021) Influence of low fines content on the liquefaction resistance of sands. European J Environ Civil Eng **0**(0): 1–20. Taylor & Francis. https://doi.org/10.1080/19648189.2021.1927195

Jradi L (2018) Study of the influence of fine particles on the properties of liquefaction of sands. Ph.D Thesis University Paris-Est

Yin K, Fauchille AL, Di Filippo E, Othmani K, Branchu S, Sciarra G, Kotronis P (2021) The influence of mixing orders on the microstructure of artificially prepared sand-clay mixtures. Adv Mater Sci Eng. https://doi.org/10.1155/2021/8552224

Zhu Z, Cheng W (2020) Parameter evaluation of exponential-form critical state line of a state-dependent sand constitutive model. Appl Sci 10(1):328. https://doi.org/10.3390/app10010328

Zhu Z, Zhang F, Dupla J, Canou J, Foerster E (2021a) Assessment of tamping-based specimen preparation methods on static liquefaction of loose silty sand. Soil Dyn Earthq Eng 143(January):106592. Elsevier Ltd. https://doi.org/10.1016/j.soildyn.2021a.106592.

Zhu Z, Zhang F, Peng Q, Chabot B, Dupla J (2021b) Development of an auto compensation system in cyclic triaxial apparatus for liquefaction analysis. Soil Dyn Earthq Eng **144**(December 2020): 106707. Elsevier Ltd. https://doi.org/10.1016/j.soildyn.2021b.106707

Chapter 4
Behaviour Under Cyclic Loading

This chapter first presents the experimental results obtained for HN31 sand specimens containing both non-plastic and plastic fines subjected to cyclic loading. Typical results are first given followed by a detailed analysis of the influence of fine particles on the development of sand liquefaction through the cyclic shear resistance curves. The mechanism behind presented in the previous chapter is here retaken to interpret the test results. The analogies existing between monotonic and cyclic loading are then established for sand-fines specimens. Finally, in order to underline the impact of loading frequency and loading shape, two specific sets of cyclic triaxial tests on clean sand specimens are respectively provided and a novel physical indicator is proposed based on the Arias intensity, allowing the combined influence of loading amplitude and shape to be simultaneously considered on the liquefaction issue.

4.1 Experimental Programme

To straightforwardly address the influence of fine particles themselves on liquefaction response, the cyclic shear resistance curves in terms of fines content F_c are here established for sand-fines specimens in a very loose state of $I_{Dmat} = 0.00$ by varying the cyclic stress ratio *CSR* from 0.120 to 0.075. Prior to shearing, all test specimens were isotropically consolidated at σ'_c equal to 200 kPa to represent the shallow liquefiable sandy deposits. Two fine particles (non-plastic C500 silica and plastic Speswhite kaolinite) were selected and the dry tamping method was thus adopted for meeting the homogeneity requirement. The detailed experimental programme for HN31-fines specimens is summarized in Table 4.1.

To further study the effect of loading frequency on liquefaction properties of sand in the present work, a series of stress-controlled cyclic triaxial tests on saturated specimens has been conducted with cyclic stress ratio $CSR = 0.220$ (CTT04-CTT08). This relatively high value was to control the test duration to be within an acceptable range. In order to directly underline the dynamic characteristics, strain-controlled

Z. Zhu, *Influence of Fine Particles on the Liquefaction Properties of a Reference Sand*, Springer Theses, https://doi.org/10.1007/978-3-031-24299-1_4

Table 4.1 Summary of the cyclic triaxial test programme for HN31-fines specimens (dry tamping method)

Reference	Type of fines	F_c (%)	*CSR*	I_{Dmat}	σ'_c (kPa)	Reconstitution method
CTTF01	C500	5.0	0.120	0.00	200	DT
CTTF02	C500	5.0	0.105	0.00	200	DT
CTTF03	C500	5.0	0.090	0.00	200	DT
CTTF04	C500	5.0	0.075	0.00	200	DT
CTTF05	C500	10.0	0.120	0.00	200	DT
CTTF06	C500	10.0	0.105	0.00	200	DT
CTTF07	C500	10.0	0.100	0.00	200	DT
CTTF08	C500	10.0	0.090	0.00	200	DT
CTTF09	C500	10.0	0.075	0.00	200	DT
CTTF10	C500	15.0	0.120	0.00	200	DT
CTTF11	C500	15.0	0.105	0.00	200	DT
CTTF12	C500	15.0	0.090	0.00	200	DT
CTTF13	C500	15.0	0.075	0.00	200	DT
CTTF14	Speswhite	5.0	0.120	0.00	200	DT
CTTF15	Speswhite	5.0	0.105	0.00	200	DT
CTTF16	Speswhite	5.0	0.090	0.00	200	DT
CTTF17	Speswhite	5.0	0.075	0.00	200	DT
CTTF18	Speswhite	10.0	0.120	0.00	200	DT
CTTF19	Speswhite	10.0	0.105	0.00	200	DT
CTTF20	Speswhite	10.0	0.090	0.00	200	DT
CTTF21	Speswhite	10.0	0.075	0.00	200	DT
CTTF22	Speswhite	15.0	0.105	0.00	200	DT
CTTF23	Speswhite	15.0	0.090	0.00	200	DT
CTTF24	Speswhite	15.0	0.075	0.00	200	DT

cyclic triaxial tests (CTT09–CTT13) over a wide range of frequencies (up to 5 Hz) were also performed. Besides, *dry sand specimens* (CTT01–CTT03) were specially involved in the test programme to explore the sliding mechanism behind of the sand grains upon shearing without the interaction of the excess pore water under different loading frequencies ($f = 0.02$–0.60 Hz). To understand the influence of loading signal shape, a series of stress-controlled cyclic triaxial tests has been also performed using three signals (triangular, sinusoidal and rectangular shape). In the case of triangular signal, three relatively high cyclic stress ratios *CSR* of 0.200, 0.220 and 0.240 were adopted so that the test duration could be controlled within an acceptable range, while for sinusoidal and rectangular signals, four ratios of 0.180, 0.200, 0.220 and 0.240 were selected, respectively. The detailed test programme for these two sets of tests is summarized in Table 4.2.

Table 4.2 Summary of the cyclic triaxial testing programme for clean sand specimens (dry tamping method)

Reference	I_{Dmat}		S_r (%)	f (Hz)	Control mode	σ'_c (kPa)	Loading shape
		CSR	Dry		Stress-controlled	100	Sinusoidal
CTT01	0.70	0.220		0.02			
CTT02				0.10			
CTT03				0.60			
CTT04	0.70	0.220	100 (saturated)	0.02			
CTT05				0.10			
CTT06				0.30			
CTT07				0.30			
CTT08				0.60			
		ε_a			Strain-controlled		
CTT09	0.70	± 0.40%	100 (saturated)	0.01			
CTT10				0.02			
CTT11				0.10			
CTT12				1.00			
CTT13				5.00			
		CSR					
CTT14	0.70	0.200	100 (saturated)	0.10	Stress-controlled	100	Triangular
CTT15		0.220					
CTT16		0.240					
CTT17	0.70	0.180					Sinusoidal
CTT18		0.200					
CTT19		0.220					
CTT20		0.240					
CTT21	0.70	0.180					Rectangular
CTT22		0.200					
CTT23		0.220					
CTT24		0.240					

4.2 Typical Test Results

4.2.1 Clean Sand Specimens

Figure 4.1 presents the results of two dense ($I_{Dmat} = 0.70$) clean sand specimens (CTT06 and CTT07) subjected to cyclic loading. The repeatability is checked here to validate the complete experimental setup and procedure, especially for this higher loading frequency $f = 0.30$ Hz. The stress–strain curves shown in Fig. 4.1a suggest that the development of the hysteresis loop was unclear at the initial stage. With further loading, the hysteresis loop became more obvious with the gradual accumulation of excess pore water pressure Δu.

The evolution of excess pore water pressure Δu versus the number of cycles N_{cyc} recorded in Fig. 4.1b shows that (i) before the excess pore water pressure ratio r_u reached a high level of 0.80, the two curves were almost identical; (ii) beyond this point, some small differences were observed owing to the fact that the hydraulic actuator could not well manage the transient variation of the axial displacement near the liquefaction failure stage, especially in the stress-controlled mode.

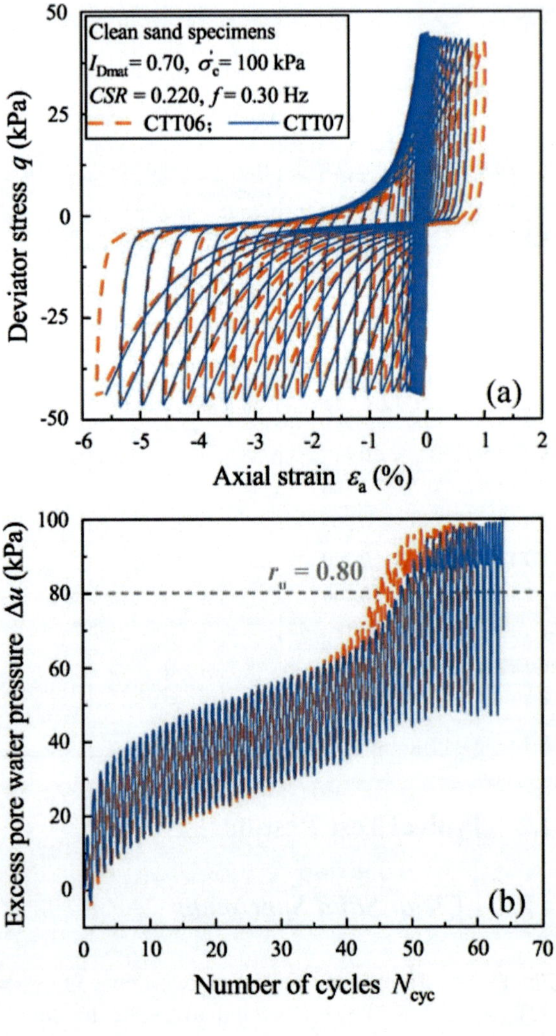

Fig. 4.1 Results obtained for clean sand specimens with $I_{Dmat} = 0.70$, $\sigma'_c = 100$ kPa and $f = 0.30$ Hz: **a** stress–strain curves; **b** excess pore water pressure versus number of cycles

4.2.2 *Sand-Fines Specimens*

Figure 4.2 presents typical test results corresponding to the development of flow liquefaction for a loose specimen ($I_{\mathrm{Dmat}} = 0.00$) of HN31 sand containing non-plastic C500 silica of $F_{\mathrm{c}} = 10.0\%$ under an isotropic consolidation stress of 200 kPa. The evolution of imposed deviator stress σ_{d} in terms of the number of cycles N_{cyc} is shown in Fig. 4.2a. Before the final failure stage, the actuator and the mechanical controller adopted in this study correctly worked and properly maintained the expected signal shape. After the triggering of sand liquefaction associated with the transient and large axial displacement, the loading system failed to perfectly manage this variation and a malfunction was indeed detected. Figure 4.2b presents the evolution of excess pore water pressure Δu with respect to N_{cyc}. As the cyclic loading was applied, Δu steadily increased at the very beginning stage. Afterwards, the growth trend became much less significant and arrived at an intermediate accumulation phase in which a quasi-linear trend was found. Finally, a sharp build-up of Δu was found near the failure stage and Δu arrived at a value very close to $\sigma'_{\mathrm{c}} = 200$ kPa. Figure 4.2c presents the relationship between the axial strain ε_{a} and N_{cyc}. At the initial stage of loading, the variation of ε_{a} was quite small. With the further development of loading and sufficient accumulation of Δu, the extent of ε_{a} was suddenly amplified at $N_{\mathrm{cyc}} = 49$ in the extension phase. In the present work, the attainment of axial strain equal to 5% in double amplitude was set as the yardstick to recognize the liquefaction failure, as mentioned previously in *Chapter* 1. Figure 4.2d presents the corresponding stress–strain curve. At the initial loading stage, the development of hysteresis loops was too subtle to be observed. However, after the triggering of liquefaction failure, the curve suddenly took the so-called "reversed-z" shape, indicating a sizable amount of energy dissipation. The corresponding effective stress path is displayed in Fig. 4.2e. The stress path started with an initial isotropic point with $\sigma'_{\mathrm{c}} = 200$ kPa and continually migrated to the left till the final failure stage. The migration of this stress path towards the original point stood for pure contractancy under cyclic loading. In addition, this stress path was clearly governed by the instability line in the extension phase, as underlined in Fig. 4.2e.

4.3 Influence of Fine Particles

In order to establish the cyclic shear resistance curve for HN31-fines specimens to assess the influence of fine particles, the experimental results obtained for different HN31-fines specimens are shown, according to the type of fine particles (C500 and Speswhite) and fines content F_{c} (5.0%, 10.0% and 15.0%), in the following.

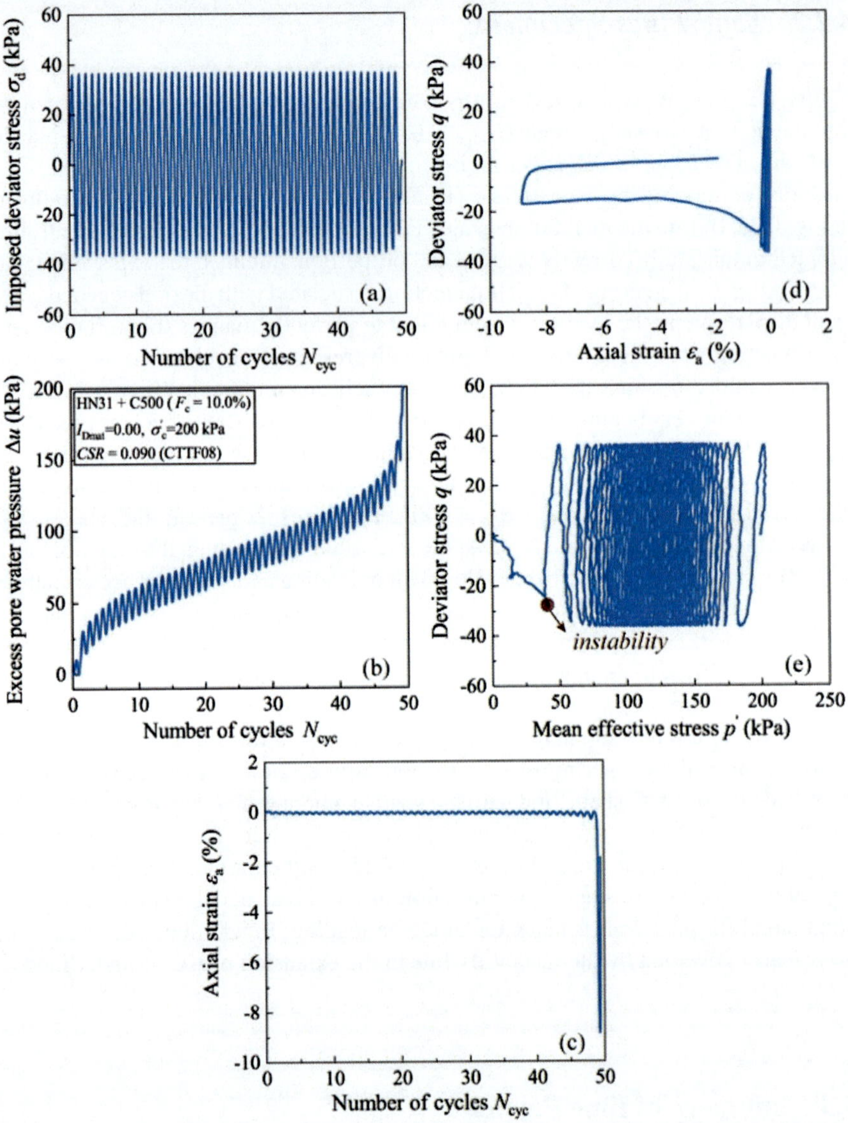

Fig. 4.2 Results obtained for HN31-C500 specimen with $I_{Dmat} = 0.00$, $\sigma'_c = 200$ kPa, $F_c = 10.0\%$: **a** imposed deviator stress; **b** excess pore water pressure versus number of cycles; **c** axial strain versus number of cycles; **d** stress–strain curve; **e** effective stress path

4.3.1 Non-plastic Fines: C500

Figure 4.3 presents the experimental results obtained for HN31-C500 specimens with $I_{Dmat} = 0.00$, $\sigma'_c = 200$ kPa and $F_c = 5.0\%$ subjected to different levels of shearing. The excess pore water pressure Δu shown in Fig. 4.3a suggests that the initial increase in *CSR* from 0.075 to 0.090 could significantly speed up the development of sand liquefaction. The number of cycles corresponding to the given failure criterion N_L decreased from 269 to 56. However, the subsequent increment in *CSR* was quite inefficient to rapidly liquefy the test specimens. The effective stress paths displayed in Fig. 4.3c were mainly governed by the instability line. A minor dilatant phase was indeed observed for $CSR = 0.075$ near the failure stage. This puzzling phenomenon might probably be attributed to the malfunction of the hydraulic actuator, as mentioned in the above section. The same conclusions can be found for HN31-C500 specimens with $F_c = 10.0\%$ in Fig. 4.4 and $F_c = 15.0\%$ in Fig. 4.5, respectively.

Figure 4.6 presents the cyclic shear resistance curves for HN31-C500 specimens corresponding to different fines contents F_c of 5.0%, 10.0% and 15.0%, respectively. In the graph, the curve clearly moves to the lower left part with the increase in F_c. This signifies that for a given amplitude of shearing, N_L needed for specimens containing more fines decreases to be much smaller than those containing fewer fines and a severe liquefaction concern shall be expected. This observation is in good agreement with that observed for HN31-C500 specimens subjected to monotonic loading (reconstituted with the dry tamping method, see Fig. 3.12 in §3.3.2). The mechanism behind explaining this decline tendency should be the same, irrespective of the loading mode (monotonic or cyclic). With the dry tamping method, some of the fine particles are located nearby the sand-sand contact points by separating them; thus, the host sand matrix yielding the global shear resistance is, to some extent, weakened and this adverse effect becomes much more important with the increase in F_c.

In order to detect the instability state of specimens containing different percentages of C500 silica, the corresponding effective stress paths are zoomed in between $p' = 0$ kPa and $p' = 100$ kPa for better precision. In Fig. 4.7, it can be seen that the instability line for each F_c can be roughly unified despite some small discrepancies due to the malfunction of the hydraulic actuator near the liquefaction stage. The relationship between the internal friction angle at the instability state ϕ^{-}_{inst} and F_c is displayed in Fig. 4.7d. A linear decreasing tendency can be found with the increase in F_c, suggesting again that the addition of C500 silica favoured the build-up of sand liquefaction.

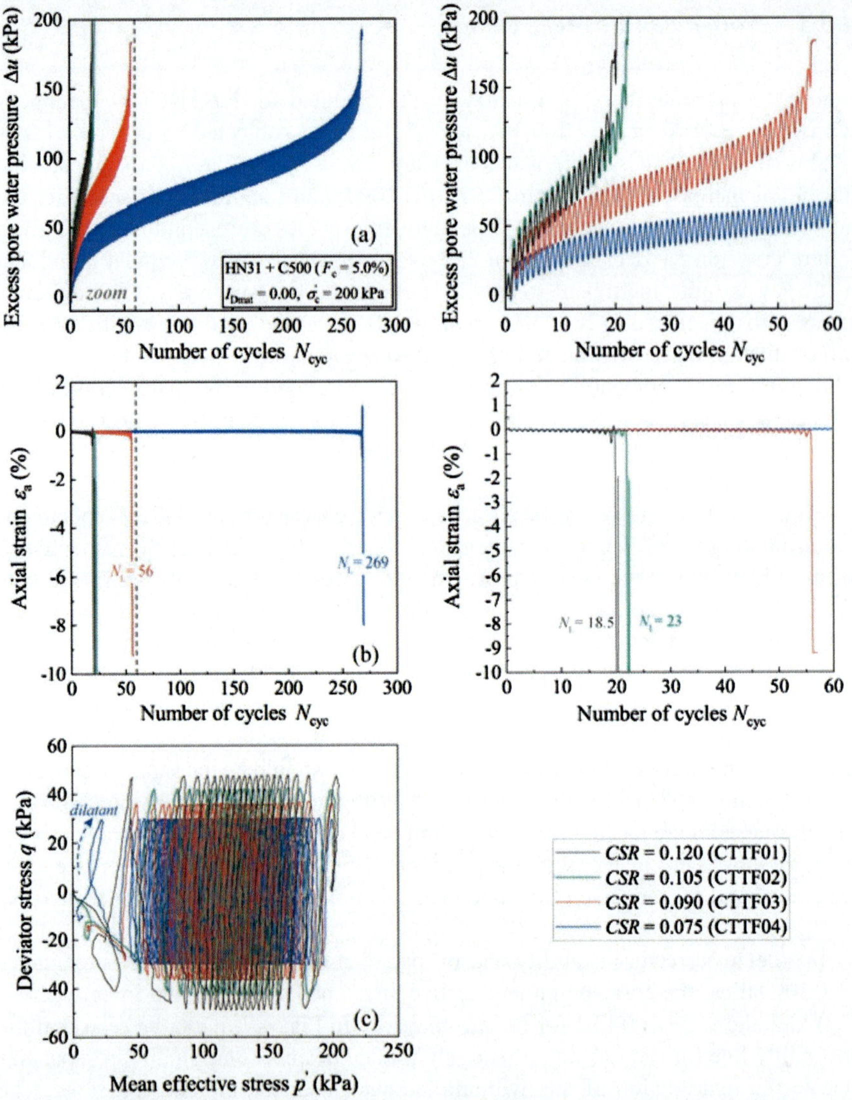

Fig. 4.3 Results obtained for HN31-C500 specimens with $I_{\mathrm{Dmat}} = 0.00$, $\sigma'_c = 200$ kPa, $F_c = 5.0\%$: **a** excess pore water pressure versus number of cycles; **b** axial strain versus number of cycles; **c** effective stress paths

4.3.2 *Plastic Fines: Speswhite*

Similarly, the experimental results for HN31-Speswhite specimens are equally provided below on the basis of F_c in order to construct the cyclic shear resistance curves to assess the influence of fines content. Regarding $F_c = 5.0\%$ in Fig. 4.8,

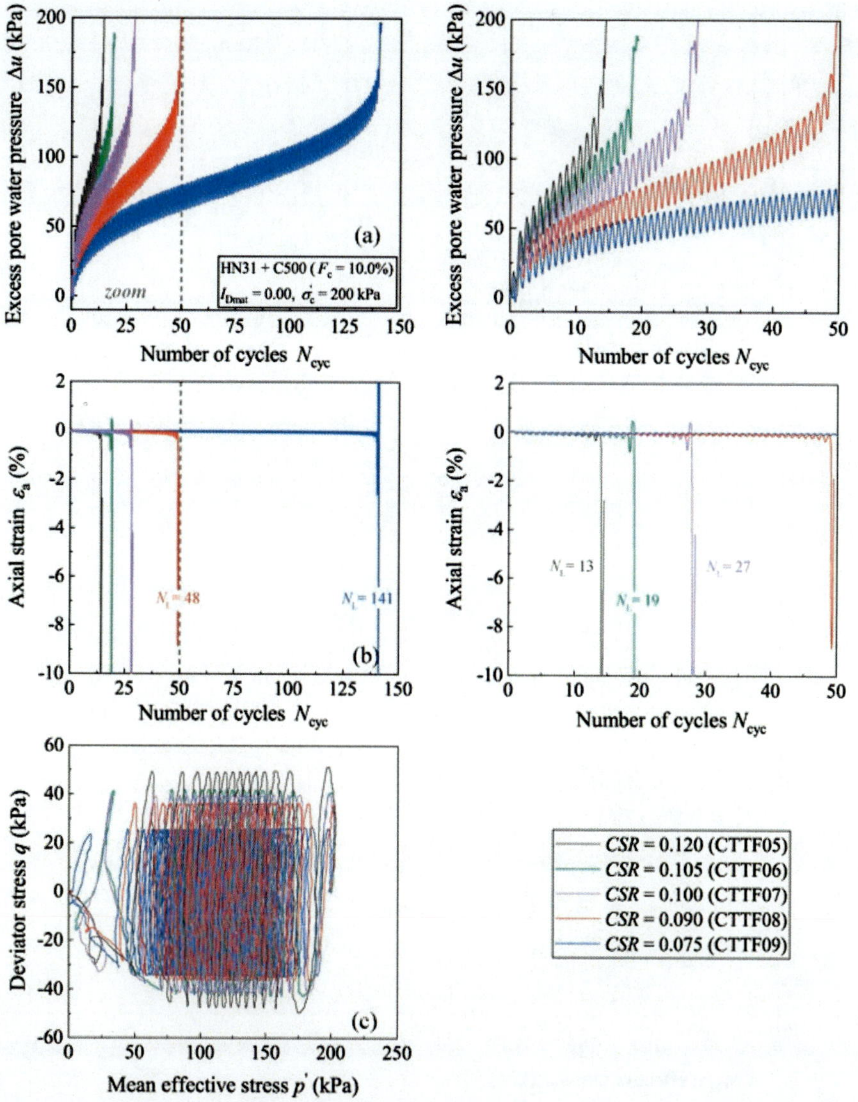

Fig. 4.4 Results obtained for HN31-C500 specimens with $I_{\text{Dmat}} = 0.00$, $\sigma'_{\text{c}} = 200$ kPa, $F_{\text{c}} = 10.0\%$: **a** excess pore water pressure versus number of cycles; **b** axial strain versus number of cycles; **c** effective stress paths

the increase in *CSR* significantly favoured the liquefaction response. The same conclusion can also be drawn from $F_{\text{c}} = 10.0\%$ in Fig. 4.9 and $F_{\text{c}} = 15.0\%$ in Fig. 4.10.

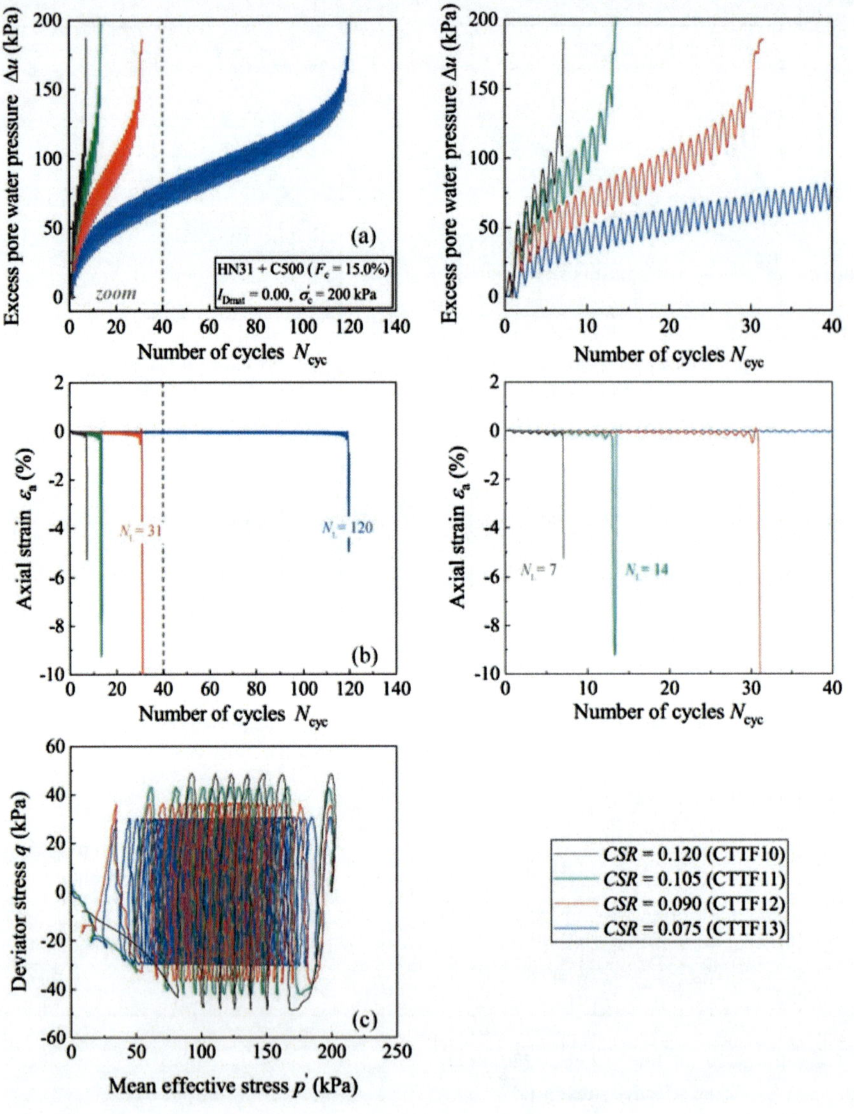

Fig. 4.5 Results obtained for HN31-C500 specimens with $I_{\text{Dmat}} = 0.00$, $\sigma'_{\text{c}} = 200$ kPa, $F_{\text{c}} =$ 15.0%: **a** excess pore water pressure versus number of cycles; **b** axial strain versus number of cycles; **c** effective stress paths

Figure 4.11 presents the corresponding cyclic shear resistance curves for HN31-Speswhite mixtures with three fines contents F_c of 5.0%, 10.0% and 15.0%, respectively. Similar to what has been previously observed for HN31-C500 specimens, the

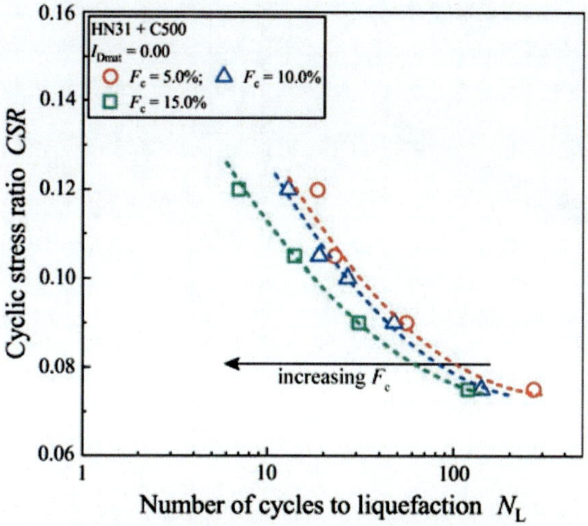

Fig. 4.6 Cyclic shear resistance curves for HN31-C500 specimens with different fines contents

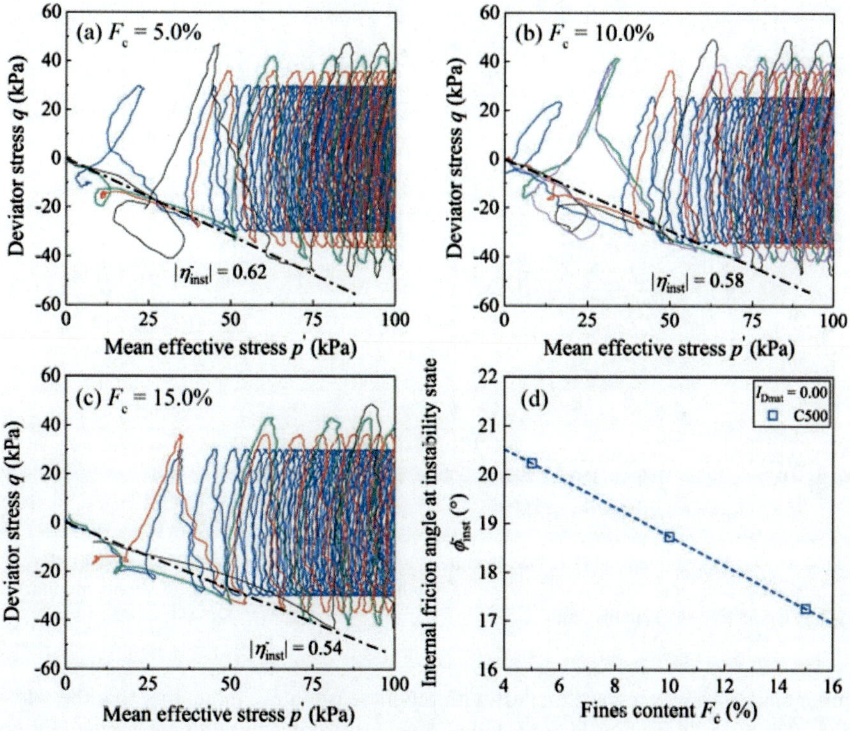

Fig. 4.7 Instability lines for HN31-C500 specimens with **a** F_c = 5.0%; **b** F_c = 10.0%; **c** F_c = 15.0%; **d** corresponding internal friction angles at the instability state

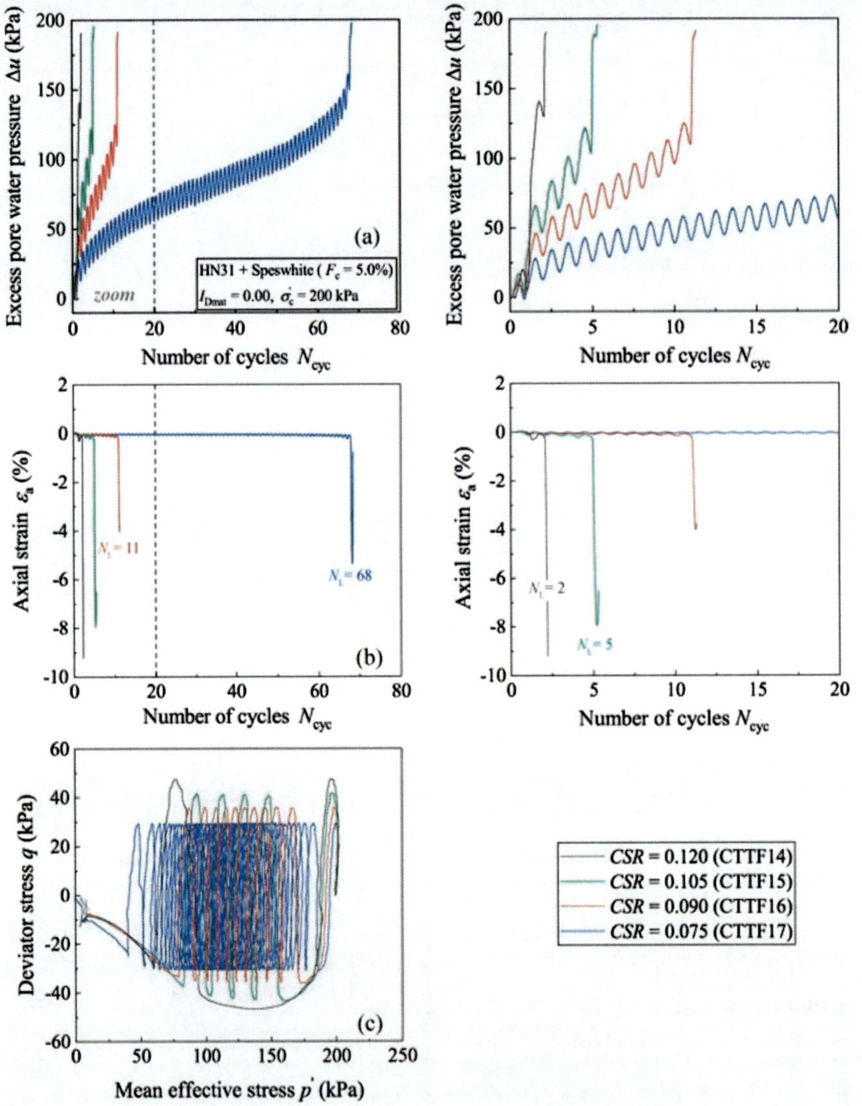

Fig. 4.8 Results obtained for HN31-Speswhite specimens with $I_{\text{Dmat}} = 0.00$, $\sigma'_{\text{c}} = 200$ kPa, $F_{\text{c}} = 5.0\%$: **a** excess pore water pressure versus number of cycles; **b** axial strain versus number of cycles; **c** effective stress paths

curve clearly moves toward the left with the increase in F_{c}, indicating that the addition of plastic Speswhite also provided a negative contribution to the liquefaction resistance.

In like manner, the corresponding effective stress paths are zoomed in between $p' = 0$ kPa and $p' = 100$ kPa to detect the instability state of specimens containing

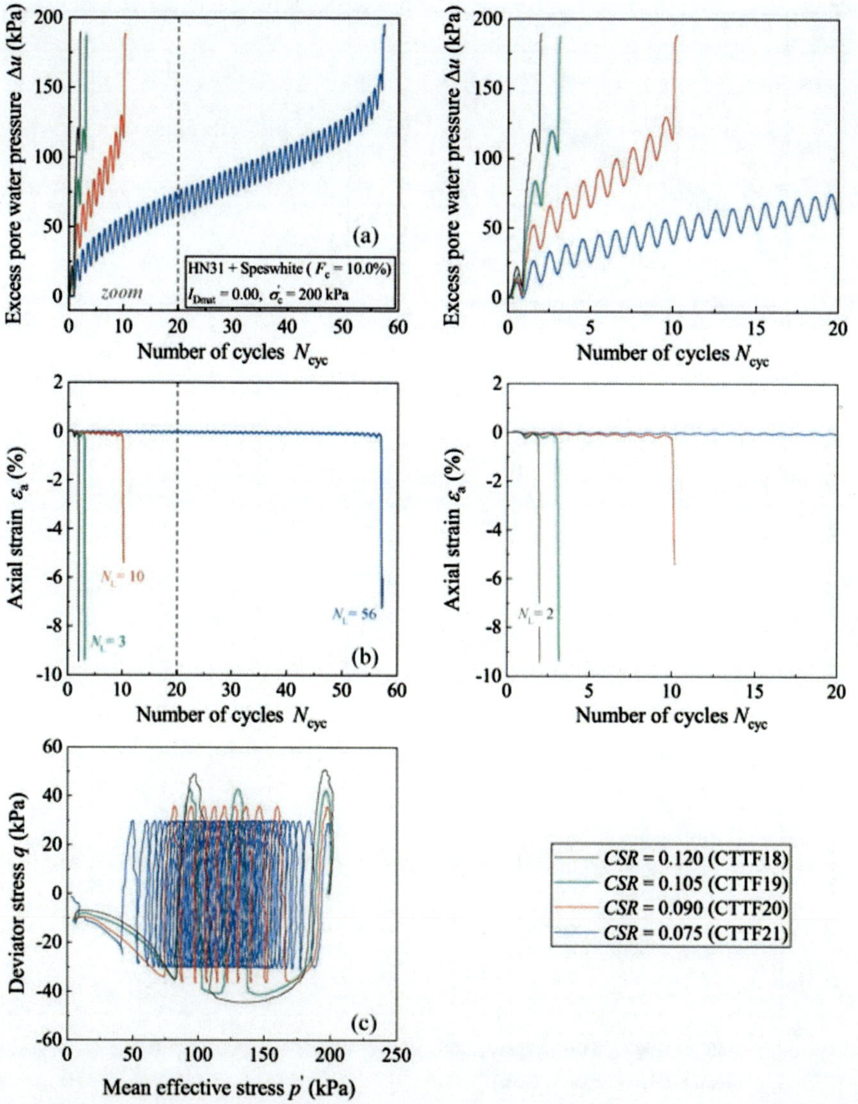

Fig. 4.9 Results obtained for HN31-Speswhite specimens with $I_{\mathrm{Dmat}} = 0.00$, $\sigma'_{\mathrm{c}} = 200$ kPa, $F_{\mathrm{c}} = 10.0\%$: **a** excess pore water pressure versus number of cycles; **b** axial strain versus number of cycles; **c** effective stress paths

different percentages of Speswhite. In Fig. 4.12, it can be seen that the instability state for each F_{c} can be roughly represented by a single straight line despite some small discrepancies. The relationship between the internal friction angle at the instability state $\phi^{-}{}_{\mathrm{inst}}$ and fines content F_{c} is presented in Fig. 4.12d for HN31-Speswhite

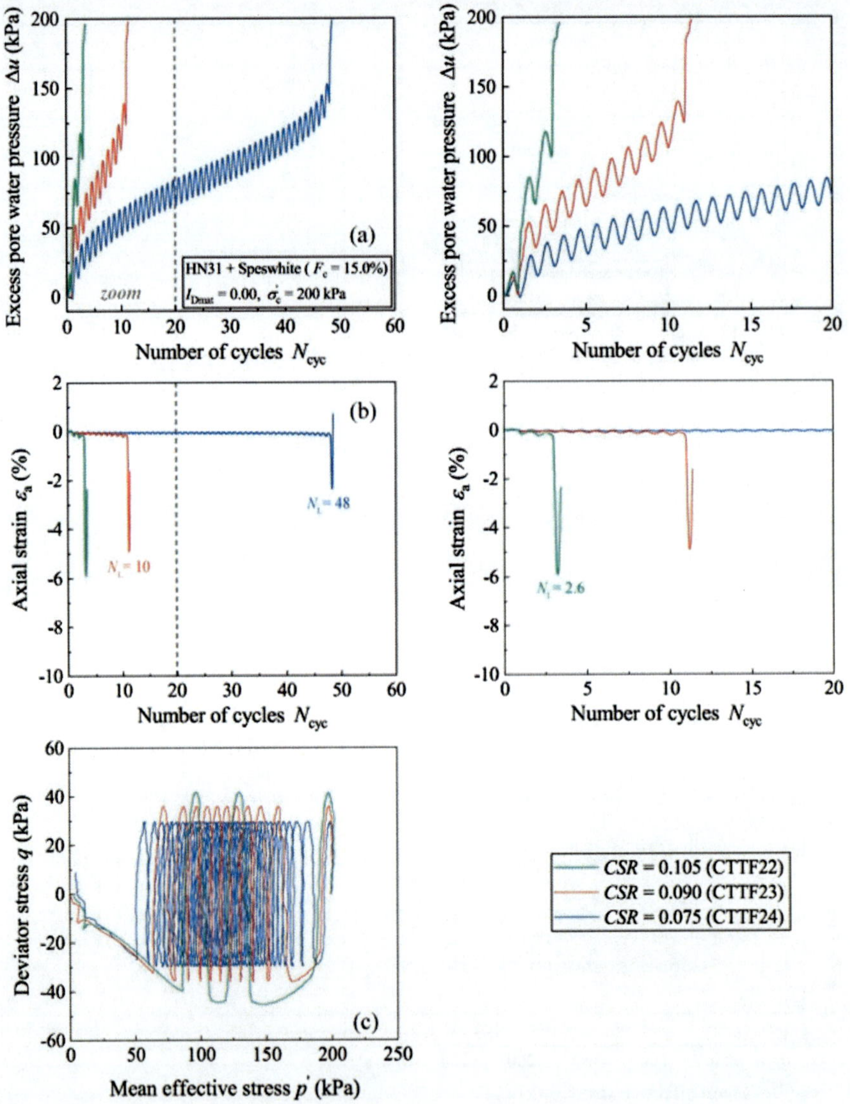

Fig. 4.10 Results obtained for HN31-Speswhite specimens with $I_{\text{Dmat}} = 0.00$, $\sigma'_c = 200$ kPa, $F_c = 15.0\%$: **a** excess pore water pressure versus number of cycles; **b** axial strain versus number of cycles; **c** effective stress paths

specimens. It can be seen that ϕ^{-}_{inst} slightly decreases with the increase in F_c, reconfirming that the addition of Speswhite in the case of the dry tamping method wakened the global resistance. For a clear exposition, ϕ^{-}_{inst} for HN31-C500 specimens drawn from Fig. 4.7 are plotted together in the graph. Regarding all three fines contents, ϕ^{-}_{inst} for HN31-C500 specimens are more important than that for HN31-Speswhite

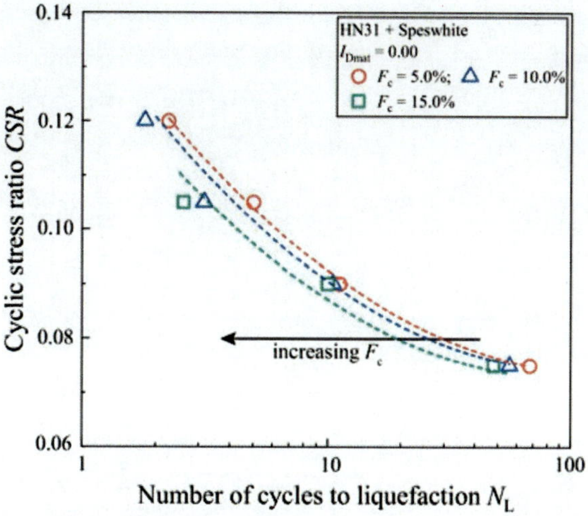

Fig. 4.11 Cyclic shear resistance curves for HN31-Speswhite specimens with different fines contents

specimens. This phenomenon reveals that the specimen containing coarser fines is much stronger than that containing finer particles for a given content.

4.3.3 Influence of Type of Fine Particles

Figure 4.13 presents the comparison of cyclic shear resistance curves between HN31-C500 and HN31-Speswhite specimens in the loose state of $I_{\mathrm{Dmat}} = 0.00$. It can be seen that for each fines content, the curve for HN31-Speswhite specimens is on the left side of that for HN31-C500 specimens, referring to a lower liquefaction resistance. As the added fines get smaller and smaller (from $D_{50}/d_{50} = 58.33$ of HN31-C500 to $D_{50}/d_{50} = 500$ of HN31-Speswhite) using the dry tamping method, the added fine particles might have more opportunities to be located nearby the sand-sand contact points by separating them. As a consequence, the finer particles such as Speswhite kaolinite are more likely to destroy the host sand matrix which yields the global shear resistance, especially as compared with the coarser grains such as C500 powder.

4.4 Comparison Between Monotonic and Cyclic Loading

4.4.1 HN31-C500 Specimens

Figure 4.14 presents the comparison between monotonic and cyclic loading on the basis of effective stress path for HN31-C500 specimens. As for $F_c = 5.0\%$ in Fig. 4.14a, two phenomena can be noticed. On the compression side, the stress path

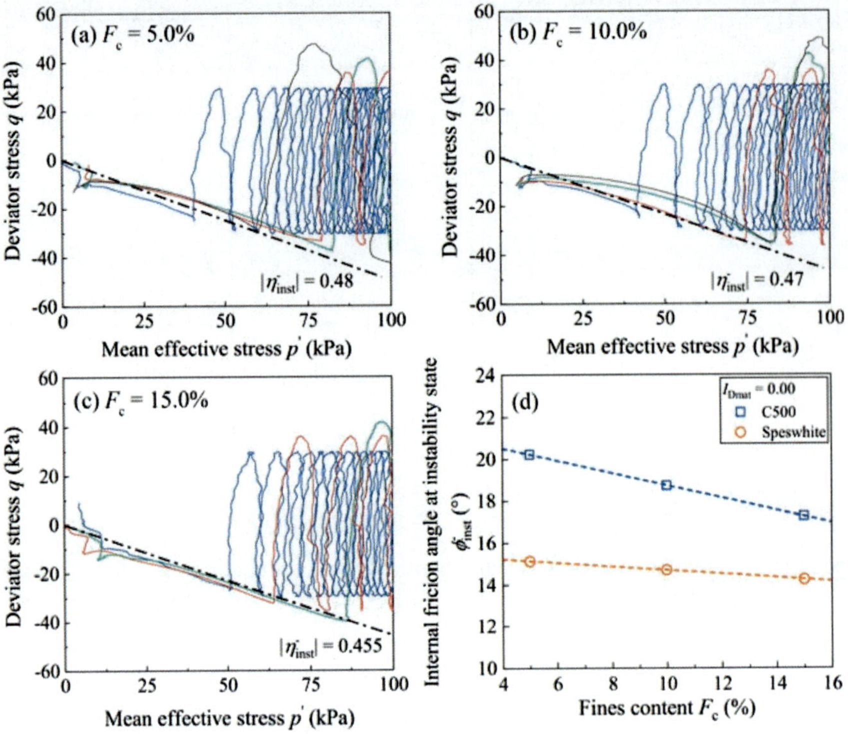

Fig. 4.12 Instability lines for HN31-Speswhite specimens with **a** $F_c = 5.0\%$; **b** $F_c = 10.0\%$; **c** $F_c = 15.0\%$; **d** corresponding friction angles mobilized at the instability state

under monotonic loading is clearly governed by the PTS line and the path steadily increases till the final failure stage. There is always a distance for the stress path under cyclic loading to touch the PTS line; as a result, a pure contractiveness is observed on this side. However, on the extension side, the path under cyclic loading finally touches, after specific cycles, the instability line whose slope value is much smaller than that of PTS. As a result, the flow liquefaction is thereafter triggered. This feature is very similar to the case (b) schematized in Fig. 1.19 (Chap. 1) and the same conclusion can also be drawn from $F_c = 10.0\%$ in Fig. 4.14b and from $F_c = 15.0\%$ in Fig. 4.14c.

4.4.2 *HN31-Speswhite Specimens*

Figure 4.15 presents the comparison between monotonic and cyclic loading on the basis of effective stress path for HN31-Speswhite specimens. As for $F_c = 5.0\%$ in Fig. 4.15a, the stress path under monotonic loading on the compression side exhibits

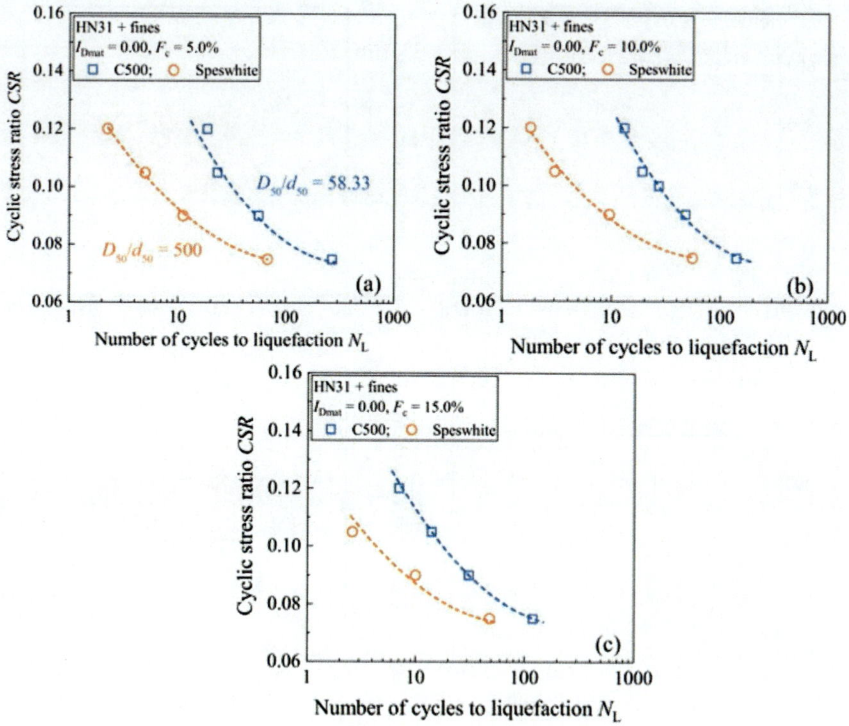

Fig. 4.13 Comparisons of cyclic shear resistance curves between HN31-C500 and HN31-Speswhite specimens at: **a** $F_c = 5.0\%$; **b** $F_c = 10.0\%$; **c** $F_c = 15.0\%$

an intermediate liquefaction behaviour. A decrease in deviator stress is observed after the initial rapid growth. And after passing through the PTS line, the specimen could regain the undrained shear resistance, corresponding to a stable dilatant behaviour. On the extension side, the stress path under cyclic loading continues to move to the left and the flow liquefaction suddenly takes place while crossing the instability line. This feature is very similar to the case (c) schematized in Fig. 1.19 (Chap. 1) and the same conclusion can also be drawn from $F_c = 10.0\%$ in Fig. 4.15b and from $F_c = 15.0\%$ in Fig. 4.15c.

4.5 Influence of Loading Frequency

Due to the inherent limitations of loading systems, cyclic triaxial tests have been primarily conducted under a lower range of loading frequencies, for instance 0.05 Hz (Benghalia et al. 2011), 0.10 Hz (Benahmed 2001; Suzuki and Yamamoto 2004; Jradi 2018), 0.20 Hz (Ghionna and Porcino 2006), 1.00 Hz (Saxena et al. 1988; Sitharam et al. 2004), etc., in order to better control the imposed loading, as well as overcome

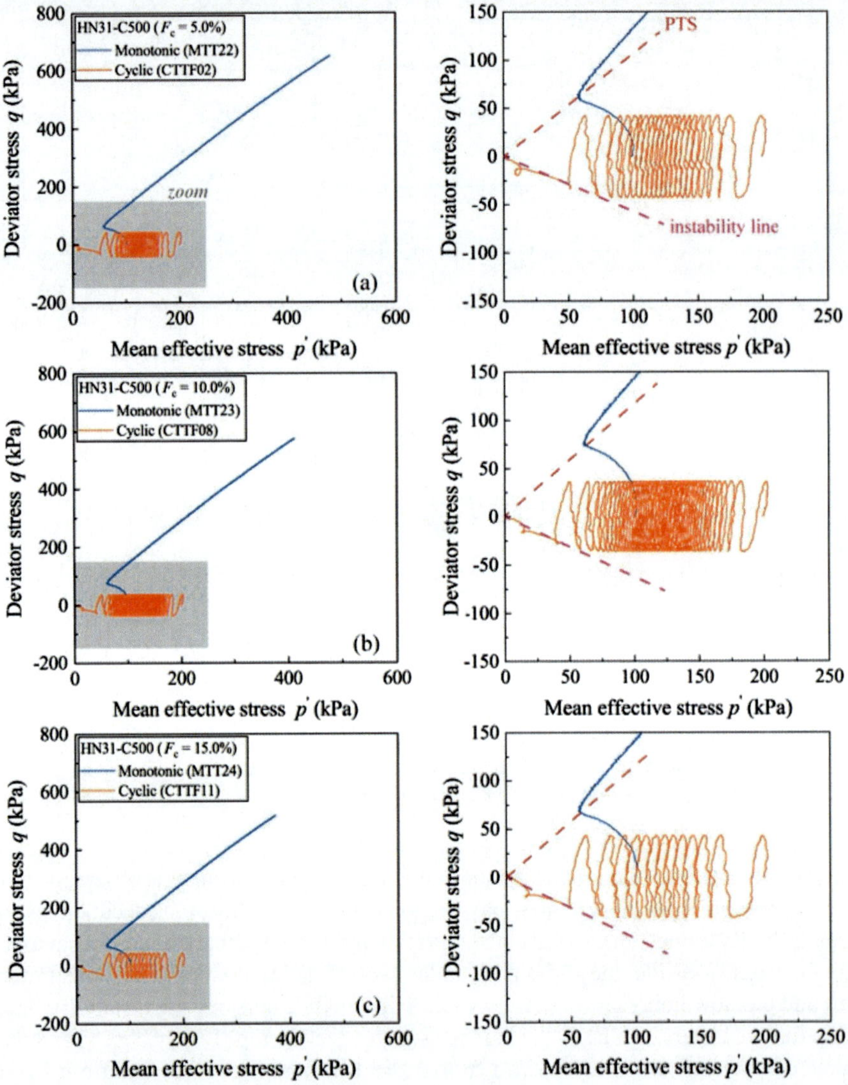

Fig. 4.14 Comparisons between monotonic and cyclic loading for HN31-C500 specimens: **a** $F_c = 5.0\%$; **b** $F_c = 10.0\%$; **c** $F_c = 15.0\%$

the measurement problems of the rapid excess pore water pressure build-up and transient axial strain. However, the frequency of a real seismic excitation typically varies in the range of a few hertz (Silva 1988; Kramer 1996; Boore and Bommer 2005), which indicates the necessity to understand the impact of the loading frequency on sand liquefaction behaviour. Considerable research has been devoted to examining this aspect over the past fifty years. However, no clear consensus has been reached, and diverse views still exist in the published literature as to whether the

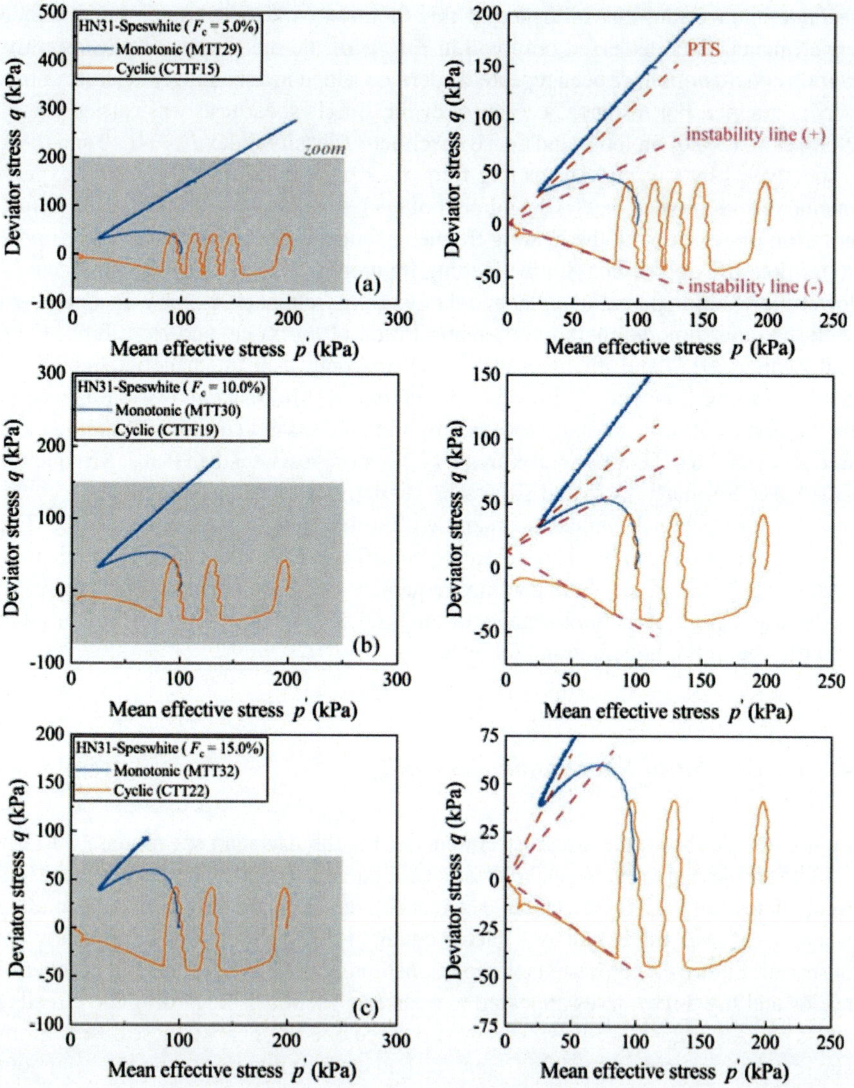

Fig. 4.15 Comparisons between monotonic and cyclic loading for HN31-Speswhite specimens: **a** $F_c = 5.0\%$; **b** $F_c = 10.0\%$; **c** $F_c = 15.0\%$

commonly used low frequency values are too conservative or not in the context of earthquake engineering. In general, the outcomes of earlier investigations revealed a negligible effect of loading frequency on liquefaction response (Peacock and Seed 1968; Yoshimi and Oh-Oka 1975; Tatsuoka et al. 1983, 1986; Polito 1999). For instance, a Roscoe-type simple shear device was used by Peacock and Seed (1968) to study the liquefaction potential of medium-dense saturated Monterey sand under different loading frequencies within the range of 0.17–4.00 Hz. However, no clear tendency concerning the influence of this factor was observed since the obtained experimental data dispersed only within ±10% of the mean value. More recently, certain conclusions have been repeatedly derived, albeit in a seemingly at least controversial manner. For instance, a series of cyclic simple shear tests was carried out by Nong et al. (2020) on loose and dense specimens (density index $I_D = 0.40$ and 0.80, respectively) by varying the loading frequency from 0.05 to 1.00 Hz. The experimental results showed that (i) the number of cycles necessary to attain a given failure criterion increased with the loading frequency and (ii) the accumulated excess pore water pressure decreased with the loading frequency. These phenomena indicated a lower liquefaction potential under a higher loading frequency, which is in agreement with the other reports, including Lee and Fitton (1969); Guo and He (2009); Feng and Zhang (2013) and Zhang et al. (2019). In contrast to this beneficial contribution of loading frequency to liquefaction resistance, Mulilis et al. (1975) seemed to be the first to observe a totally reversed trend that the sand specimen examined at a frequency of 0.017 Hz was approximately 20% more resistant than that at a frequency of 1.00 Hz. Similarly, Dash and Sithara (2016) demonstrated the increase of the cyclic resistance of original Ahmadabad sand with a decrease in loading frequency. The references mentioned above are summarized in Table 3.1 and it can be noted that the divergence in the effect of the loading frequency seems not to be possibly attributed to the employed test method, density of the sand specimen or failure criteria used to identify the initial liquefaction (Table 4.3).

4.5.1 *Dry Sand Specimens*

Figure 4.16A shows the stress–strain curves for the dry sand specimens (CTT01-CTT03) subjected to cyclic shearing of a *CSR* equal to 0.220 under different loading frequencies f of 0.02, 0.10 and 0.60 Hz, respectively. Despite the fact that the loading frequency f was increased by a factor equal to 30, only minor differences were observed. Figure 4.16b presents the relationship between axial strain and number of cycles and the three curves appeared to be nearly identical. Regarding the effective stress paths in Fig. 4.16c, three curves overlapped one another. All these observations suggest that the applied loading frequency had a negligible impact on the behaviour of dry sand specimens.

4.5.2 *Saturated Sand Specimens*

The experimental results concerning saturated HN31 sand specimens subjected to alternated shearing of *CSR* equal to 0.220 under different loading frequencies (CTT04-CTT08) are regrouped in the subplots of Fig. 4.17. The imposed deviator stress σ_d acting on the test specimens sketched in Fig. 4.17a suggests that prior to the triggering of liquefaction, the hydraulic actuator remained able to generate the

Table 4.3 Summary of literature review pertaining to the loading frequency effect

No	Reference	Sand	Test method	σ'_c (kPa)	I_D	Failure criteria	f (Hz)	Effect
1	Peacock and Seed (1968)	Monterey	SST [(1)]	500	0.5	$r_u = 1$	0.17–4	Minor
2	Yoshimi and Oh-Oka (1975)	Bandaijima	RTT [(2)]	95.2	≈ 0.4	$r_u = 1$	1–12	
3	Tatsuoka et al. (1983)	Toyoura	CTT [(3)]	98	0.50–0.80	$\varepsilon_a =$ 10% (DA [(4)])	0.05, 0.5	
4	Tatsuoka et al. (1986)	Toyoura	CTT	100	0.5	$\varepsilon_a =$ 10% (DA)	0.05–1	
5	Polito (1999)	Monterey	CTT	100	0.74	$r_u = 1$	0.5, 1	
6	Lee and Fitton (1969)	EI Monte and Silt	CTT	103	0.50, 0.75	$\varepsilon_a =$ 5%, 10% or 20% (DA)	0.17–1	More stable at higher f
7	Guo and He (2009)	NA [(5)]	CTT	100–300	0.28, 0.70	$r_u = 1$	0.05–1	
8	Feng and Zhang (2013)	Fujian sand	CTT	50	0.3	$r_u = 1$	0.05–2	
9	Zhang et al. (2015)	NA	CTT	100	0.65	$r_u = 1$	0.5–2	
10	Zhang et al. (2019)	Wuchuan sand	CTT	50, 100, 150	0.55	$r_u = 1$	1–3	
11	Nong et al. (2020)	Nakdong River sand	SST	50, 100, 200	0.40, 0.80	$\varepsilon_a =$ 7.5% (DA)	0.05–1	
12	Mulilis et al. (1975)	Monterey	CTT	100	0.5	$\varepsilon_a =$ 5% (DA)	0.017–1	More stable at lower f
13	Dash and Sithara (2016)	Original Ahmadabad sand	CTT	100	0.54	$r_u = 1$	0.1–0.5	

Note (1) SST: simple shear test; (2) RTT: ring torsion test; (3) CTT: cyclic triaxial test; (4) DA: double amplitude; (5) NA: not available

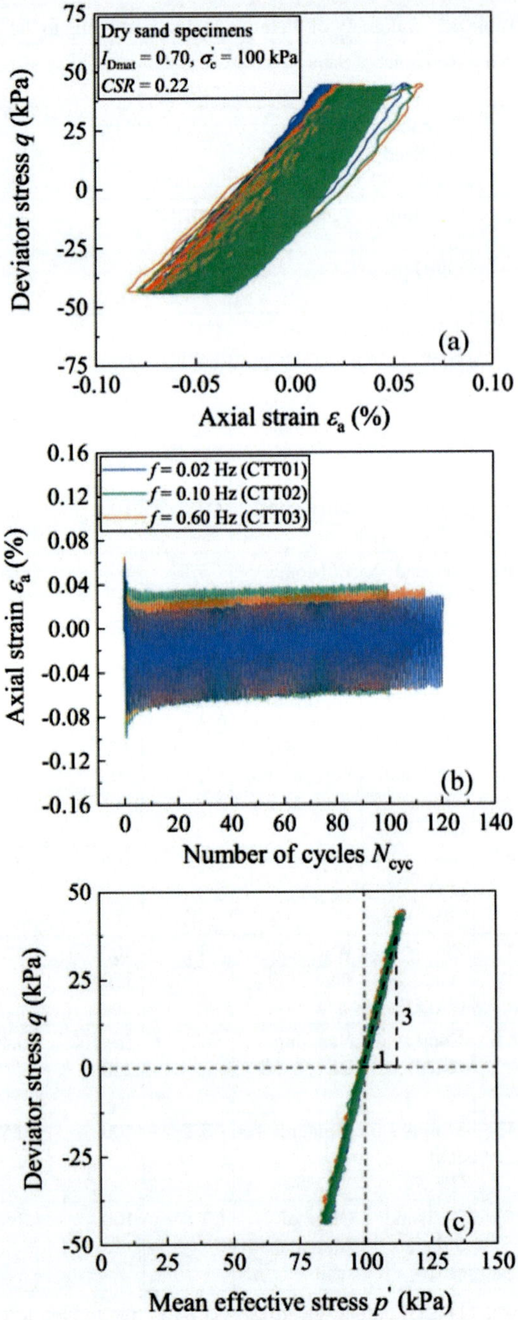

Fig. 4.16 Results obtained for dry sand specimens in stress-controlled mode with $I_{Dmat} = 0.70$, $\sigma'_c = 100$ kPa, $CSR = 0.22$ under different loading frequencies: **a** stress–strain curves; **b** axial strain versus number of cycles; **c** effective stress paths

desired loading with high robustness in the target range of loading frequencies (from $f = 0.02$ to 0.60 Hz).

The relationship between Δu and N_{cyc} depicted in Fig. 4.17b shows that (i) the "double peak mechanism" related to cyclic mobility phenomenon for the dense sand specimen became increasingly notable while approaching $r_{\mathrm{u}} = 1.0$; (ii) the applied loading frequencies played a very significant role on the response of saturated sand specimens, especially as compared with the dry sand specimens; (iii) for a given N_{cyc}, Δu significantly increased as the loading frequency f decreased.

The relationship between ε_{a} and N_{cyc} depicted in Fig. 4.17c suggests that (i) the observed axial strain range of saturated specimens was much larger than that of the dry sand specimens; (ii) in the initial phase of loading ($N_{\mathrm{cyc}} < 10$), ε_{a} did not exhibit considerable fluctuations with the variation in the loading frequency f; (iii) nevertheless, the difference became increasingly important with further loading, particularly with the development of transient deformation near the liquefaction stage.

Figure 4.18 presents the effective stress paths for the same tests (CTT04-CTT08). In the graph, the N_{cyc} required to reach the cyclic mobility increased with increasing f, reconfirming that sand liquefaction resistance indeed increased with the increase of loading frequency. It should be mentioned that in this stress-controlled mode, the post-liquefaction behaviour could not be well controlled due to the following two facts: (i) for higher loading frequencies, the loading piston was required to very rapidly reverse the loading direction from triaxial compression to extension and vice versa and it normally failed to perfectly manage this transient displacement with the occurrence of liquefaction collapse; (ii) the rapid movement of the loading piston certainly resulted in a possible variation of confining pressure provided by a single air pressure regulator. The above reasoning could explain why two distorted "butterfly wings" appear in Fig. 4.18c–d, especially for higher loading frequencies of $f = 0.30$ and 0. 60 Hz.

Figure 4.19 presents the experimental results obtained for saturated sand specimens subjected to alternated axial strain with $\varepsilon_{\mathrm{a}} = \pm 0.40\%$ (CTT09–CTT13) under different loading frequencies varying from $f = 0.01$–5.00 Hz. The stress–strain curves in Fig. 4.19a show that the actuator functioned correctly as all axial strains fell within the prescribed range in relation to the imposed loading despite the significant increase in f. In terms of deviator stress q, the amplitude value was much greater on the compression side than that on the extension side. This asymmetrical property is in accordance with the experimental observations reported in the literature (Kumar et al. 2017). The relationship between Δu and N_{cyc} depicted in Fig. 4.19b shows that (i) with the first small increase in f (from 0.01 Hz to 0.02 Hz), two curves almost coincided with each other and no clear tendency upon f could be found; (ii) with the further increase, the build-up of Δu decreased to be smaller with increasing f. For strain-controlled cyclic triaxial tests, r_{u} is commonly taken as a reasonable yardstick to identify the onset of sand liquefaction, and a value of $r_{\mathrm{u}} = 0.90$ is here thought. It can be seen from Fig. 4.19c that number of cycles required to the attainment of this considered criterion N_{L} increased with the increasing f, which is consistent with the observations in stress-controlled tests that a higher loading frequency can

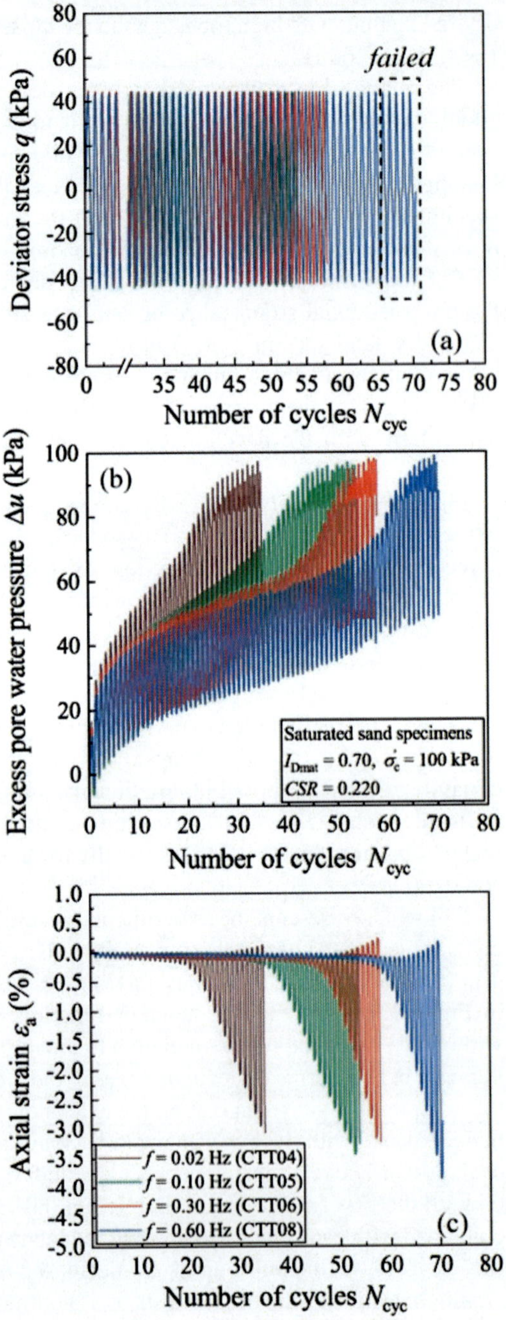

Fig. 4.17 Results obtained for saturated sand specimens in stress-controlled mode with $I_{Dmat} = 0.70$, $\sigma'_c =$ 100 kPa, $CSR = 0.220$ under different loading frequencies: **a** loading signal; **b** excess pore water pressure versus number of cycles; **c** axial strain versus number of cycles

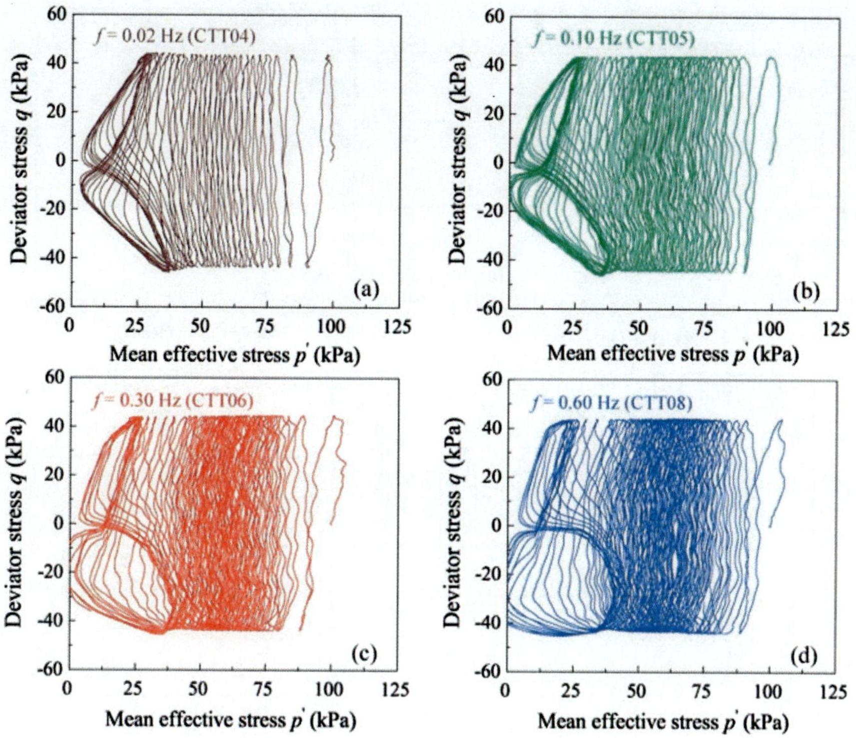

Fig. 4.18 Effective stress paths for saturated sand specimens in force-controlled mode with $I_{\mathrm{Dmat}} = 0.70$, $\sigma'_c = 100$ kPa, $CSR = 0.220$ under different loading frequencies

suppress the progress of sand liquefaction. In fact, the above experimental outcomes are consistent with the conclusion drawn by Ishihara (1995) and Xie (2011) that the resistance of soil to deformation would be greater under loading with a higher frequency than that for the loading with a lower frequency.

For granular materials such as sand, the overall soil resistance to deformation (e.g., axial strain developed in a cyclic triaxial test) upon external loading beyond the initial elastic range is mainly the result of the irreversible sliding of the sand grains against one another, which leads to the generation of the occlusal friction resulting from the reciprocal constraint of the adjacent sand grains on their relative movements. From a microscopic viewpoint, this occlusal friction between sand grains could be determined considering the following two factors according to the friction law (friction $f = N \times \mu$): (i) the normal force N perpendicular to the contact point standing for the compaction degree of the sand grains, which can be globally represented by the consolidation pressure acting on the test specimen; (ii) the frictional coefficient, μ, which can be represented by the intrinsic frictional angle of the granular material (HN31 sand in the present work). Thus, it can be logically deduced that for sand specimens composed of a given sand under a certain level of consolidation pressure, the occlusal friction against external loading can be assumed to be constant as long

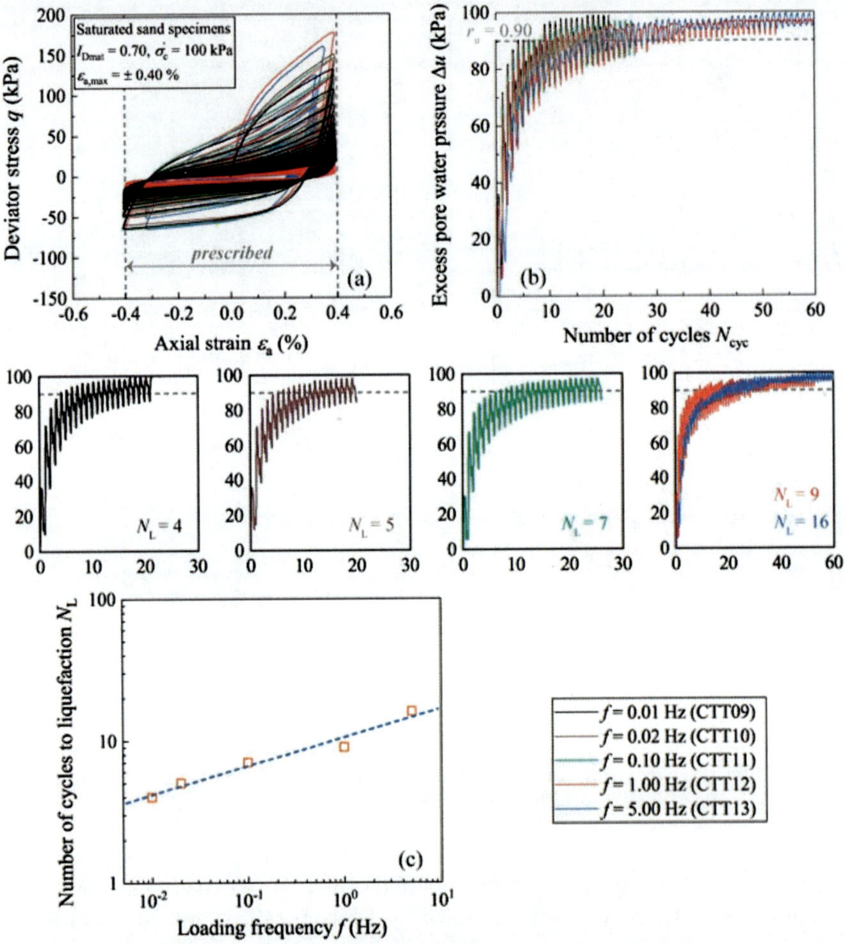

Fig. 4.19 Results obtained for saturated sand specimens in displacement-controlled mode with $I_{Dmat} = 0.70$, $\sigma'_c = 100$ kPa, $\varepsilon_a = \pm 0.40\%$ under different loading frequencies

as relative sliding occurs. And this occlusal friction is further independent of the deformation rate. This mechanism behind could explain why very similar responses were observed in the case of dry sand specimens (CTT01-CTT03).

However, this argument could not legitimately explain the effect of loading frequency on the response of the saturated sand specimens. As the time interval at which deformation occurs becomes very short in cyclic triaxial tests, it is to be borne in mind that the inertia force of sand grains themselves is another kind of agency that needs to be critically taken into account in the range of the transient deformation associated with sand liquefaction. In general, the inertia force that characterizes the dynamic behaviour as distinguished from the static one depends on the (i) mass m of moving sand grains and (ii) acceleration a linked to the strain rate of the considered sand grains. Figure 4.20 schematizes two conceptual models of the inertia force appearing in extension and compression half cycles, respectively. In the axial direction and the case of the extension cycles depicted in Fig. 4.20a, the inertia force appears opposite to the sand grains' movement, thereby enhancing the axial resistance to the applied axial loading. In the case of the compression cycles, as shown in Fig. 4.20b, the presence of inertia force equally makes a beneficial contribution to the axial resistance in the same manner. In other words, when the strain rate applied to the test specimen becomes conspicuously important with the appearance of transient deformation while approaching or attaining the sand liquefaction, *the test specimen cannot experience a sufficient time lag, allowing the full build-up of sand liquefaction to be achieved since the applied loading rapidly reverses its direction in the case of higher loading frequencies.* This rationale behind is capable to inhibit the accumulated development of $\varepsilon_a/\Delta u$ in stress/strain-controlled tests and might reasonably explain why a remarkable effect of the loading frequency occurred in saturated sand specimens subjected to alternated shearing.

For the horizontal direction in the extension and compression cycles, two distinct mechanisms can be identified: (i) the inertia force exists in the form of weakening the lateral confinement on the extension side due to the necking pattern of deformation (see Fig. 4.20a) and (ii) the inertia force still makes a positive contribution to the lateral confinement on the compression side due to the barrel pattern of deformation (see Fig. 4.20b). Therefore, the enhanced confinement on the compression side makes the relative sliding between adjacent grains much harder, yielding a better resistance to axial deformation. This mechanism behind could logically explain why the sand specimen has a clear tendency to accumulate more axial strain on the extension side than that on the compression side, especially after the triggering of sand liquefaction.

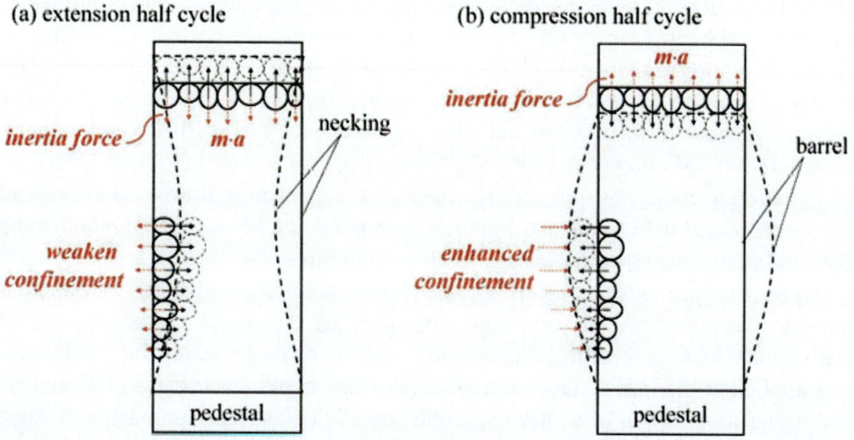

Fig. 4.20 Schematic of the inertia force on sand grains: **a** case of extension half cycles; **b** case of compression half cycles (after Omidvar et al. 2012)

4.6 Influence of Loading Signal Shape

Due to the fact that a sinusoidal loading is characterized by a single component (determined by its circular frequency) in the frequency domain, the sinusoidal loading commonly employed under laboratory conditions is far from representing a real seismic excitation, which is typically composed of a variety of frequential components covering a range of a few hertz. Transformation of the loading shape could directly reconstitute the frequential components (e.g., sinusoidal to triangular or rectangular shape). This aspect has not been well reported in the literature, likely because of the difficulty in correctly creating different loading shapes in the laboratory. With this in mind, the section first presents a set of cyclic triaxial tests (see Table 4.2) conducted on clean sand specimens with the use of three different shapes: triangular, sinusoidal and rectangular loadings, to assess the corresponding liquefaction response.

Finally, to obtain a unified sand liquefaction response, a new parameter is proposed in what follows on the basis of the Arias intensity, allowing the coupled influence of the amplitude and loading shape to be simultaneously considered.

4.6.1 Experimental Results

Figure 4.21 presents the relationship between the recorded imposed deviator stress and the number of cycles N_{cyc} using three loading shapes, respectively. The generated triangular and sinusoidal shapes displayed in Fig. 4.21a, b were relatively smooth, indicating that the actuator worked correctly. In the case of rectangular shape, some small fluctuations can be observed especially in the first half cycle (as outlined in Fig. 4.21c), which was probably due to the limitations of the applied PID controller. However, in the subsequent loading, a proper rectangular form could be well maintained. Figure 4.22 shows the theoretical and actual generated loadings for a given cyclic stress ratio $CSR = 0.220$ and a loading frequency $f = 0.10$ Hz. In the graph, some minor time lags appear, especially for the first cycles since the applied mechanical controller could not immediately manage the error in real-time. However, all curves respected well to the amplitude value related to $CSR = 0.220$, reconfirming the good performance of the triaxial setup used in this work.

Figure 4.23 presents the results obtained for the specimens subjected to triangular loading. The evolution of excess pore water pressure Δu versus N_{cyc} shows that all Δu steadily increased with the increasing number of cycles as cyclic axial stress was applied at the initial stage, and the rate of this trend became less conspicuous due to the progression of cycles. Near the end of the test, Δu approached a value almost equal to the consolidation pressure of 100 kPa, and the so-called "double peak mechanism" became increasingly notable at the same time indicating the occurrence of cyclic mobility, which is consistent with $I_{Dmat} = 0.70$. It should be noted that the increase in CSR significantly reduced the N_{cyc} required to attain the above stage. The

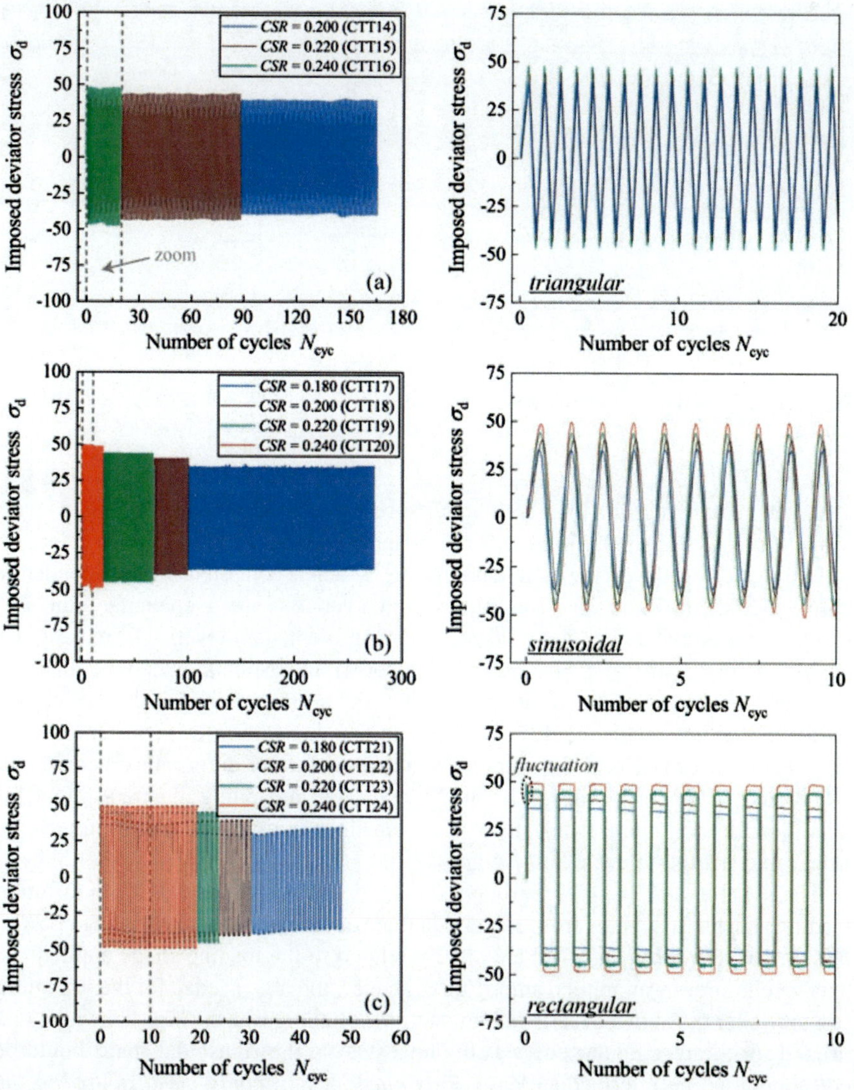

Fig. 4.21 Evolutions of imposed deviator stress in terms of number of cycles: **a** triangular loading; **b** sinusoidal loading; **c** rectangular loading

evolution of axial strain ε_a versus N_{cyc} is plotted in Fig. 4.23b. The graph indicates that in the early stage of loading, the variation in ε_a was nearly flat and only slightly tilted towards the extension side. After the coupled Δu accumulated to a certain level of approximately 70 kPa, the extent of ε_a was sharply amplified and ε_a mostly developed on the extension side. Similar experimental outcomes could also be observed in the case of sinusoidal loading, as shown in Fig. 4.24.

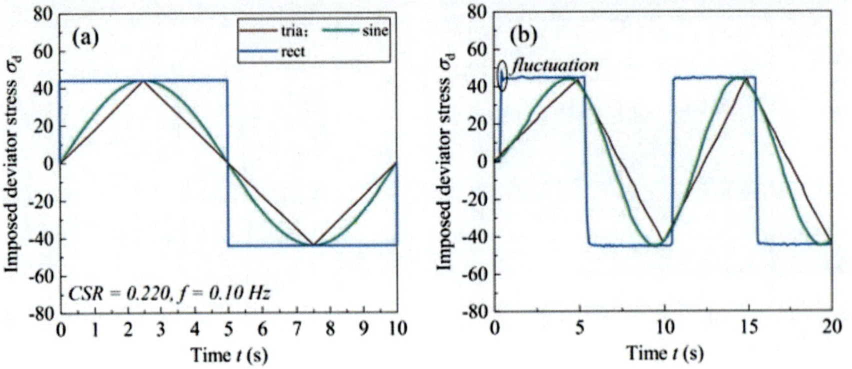

Fig. 4.22 **a** Theoretical loadings; **b** actual generated curves with $CSR = 0.220$ and $f = 0.10$ Hz ($T = 10$ s)

As for rectangular loading, the overall specimens' responses of Δu and ε_a with respect to N_{cyc} are shown in Fig. 4.25a, b and it can be seen that the liquefaction softening increased with the increment in *CSR*, which is consistent with the tendency observed in the former cases. The local enlarged versions are given on the right. The graph suggests that a huge ε_a could occur during the instant reversal of rectangular loading (compression to extension and vice versa). In particular, when the imposed deviator stress σ_d remained at its peak value, the corresponding ε_a that developed during the loading reverse could be well maintained. In addition, this constant loading stage continued to deform the specimens and some further deformations could still be produced. This phenomenon became much more prevailing in extension cycles than compression ones, especially while sand liquefaction was approaching (e.g., from 5th cycle to 9th cycle in the graph).

The subplots of Fig. 4.26 regroup the evolution of ε_a versus N_{cyc} with different loading shapes. It appears from Fig. 4.26a that for a given level of shearing (*CSR* = *0.200*), the responses observed are clearly related to the loading shape applied. As axial cyclic stress was applied among three shapes, the N_{cyc} needed for the triangular one to trigger the above-described liquefaction softening was the highest, whilst that needed for the rectangular one was the lowest. And the sinusoidal shape could be considered as the intermediate case. *This phenomenon could be attributed to the greater imposed loading acting for a longer duration* (e.g., rectangular loading), and as a result, a higher level of specimen deformation and liquefaction softening could be expected. In addition, this mechanism is also thought to be related to the inertial effect of clean sand (time-dependent behaviour), as explained in the foregoing section.

Figure 4.27 summarizes the evolution of the number of cycles to liquefaction N_L versus *CSR*. The graph suggests that for a given failure criterion (e.g., $\varepsilon_a = 5\%$), each loading shape created its own cyclic strength curve for the specimens even having the same testing parameters. In addition, the difference between triangular and sinusoidal shapes was relatively small, and the two curves were almost parallel to each

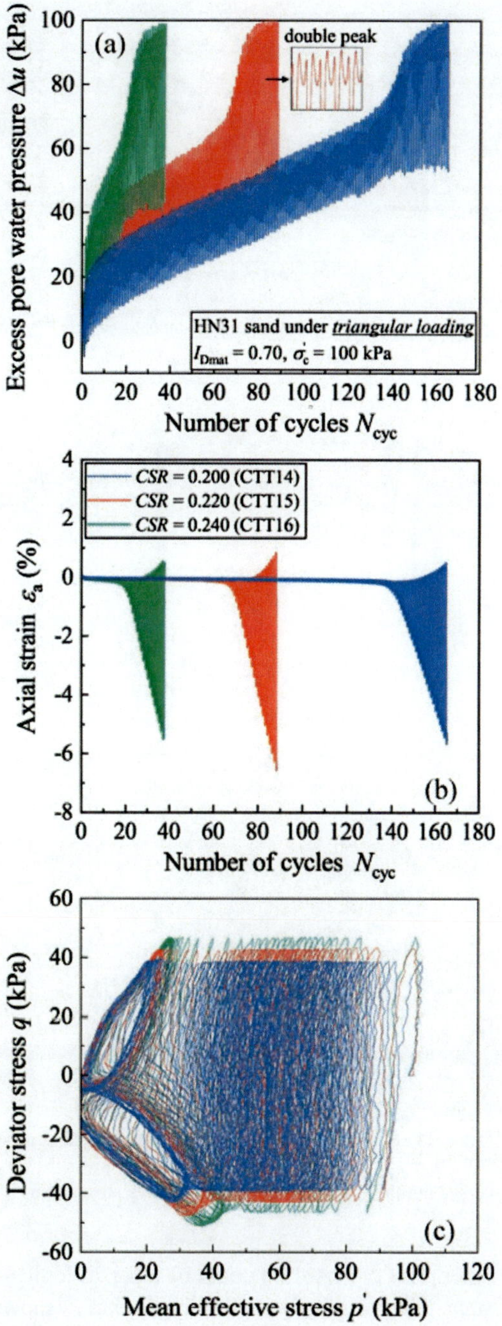

Fig. 4.23 Results obtained for HN31 sand specimens subjected to triangular loading with $I_{\mathrm{Dmat}} = 0.70$, $\sigma'_{\mathrm{c}} = 100$ kPa: **a** excess pore water pressure versus number of cycles curves; **b** axial strain versus number of cycles curve; **c** effective stress paths

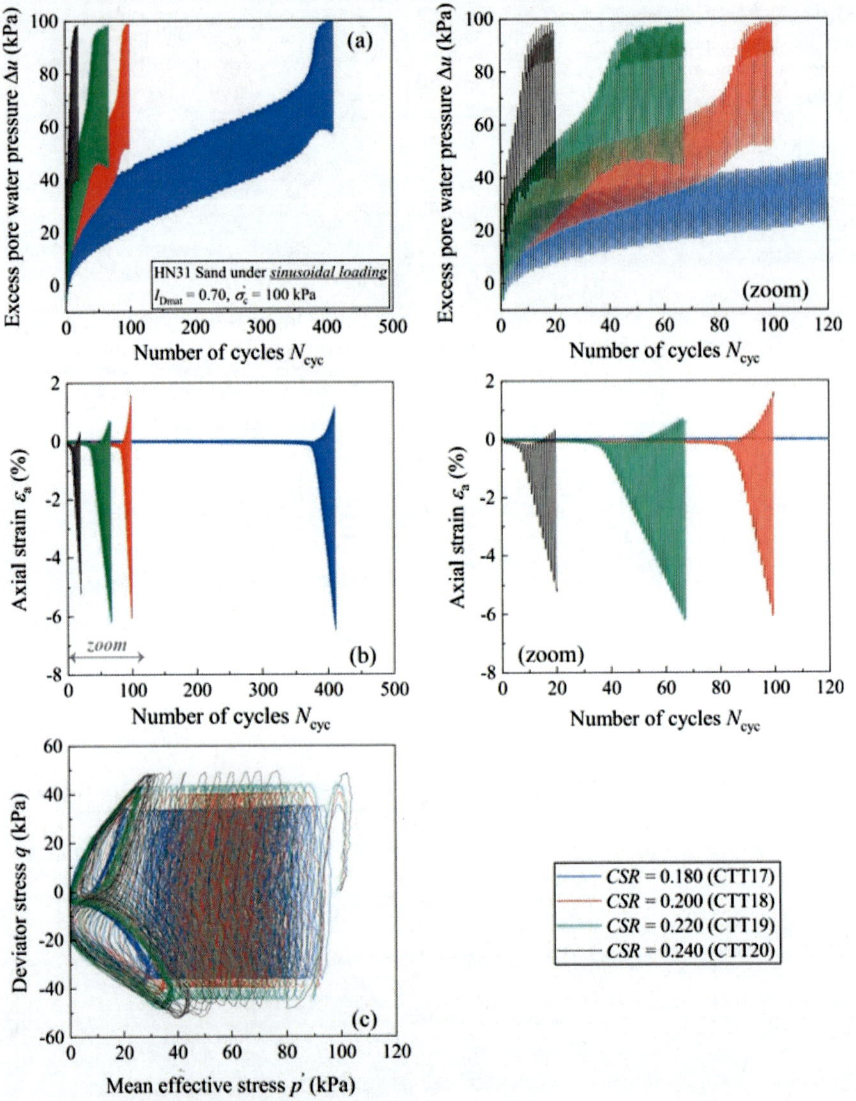

Fig. 4.24 Results obtained for HN31 sand specimens subjected to sinusoidal loading with $I_{\mathrm{Dmat}} = 0.70$, $\sigma'_c = 100$ kPa: **a** excess pore water pressure versus number of cycles curves; **b** axial strain versus number of cycles curve; **c** effective stress paths

other. This observation could be taken as logical because the imposed triangular and sinusoidal loadings were highly similar, as shown in Figs. 4.21 and 4.22. In contrast, the curve for rectangular shape is visibly located in the lowest part of the graph. For

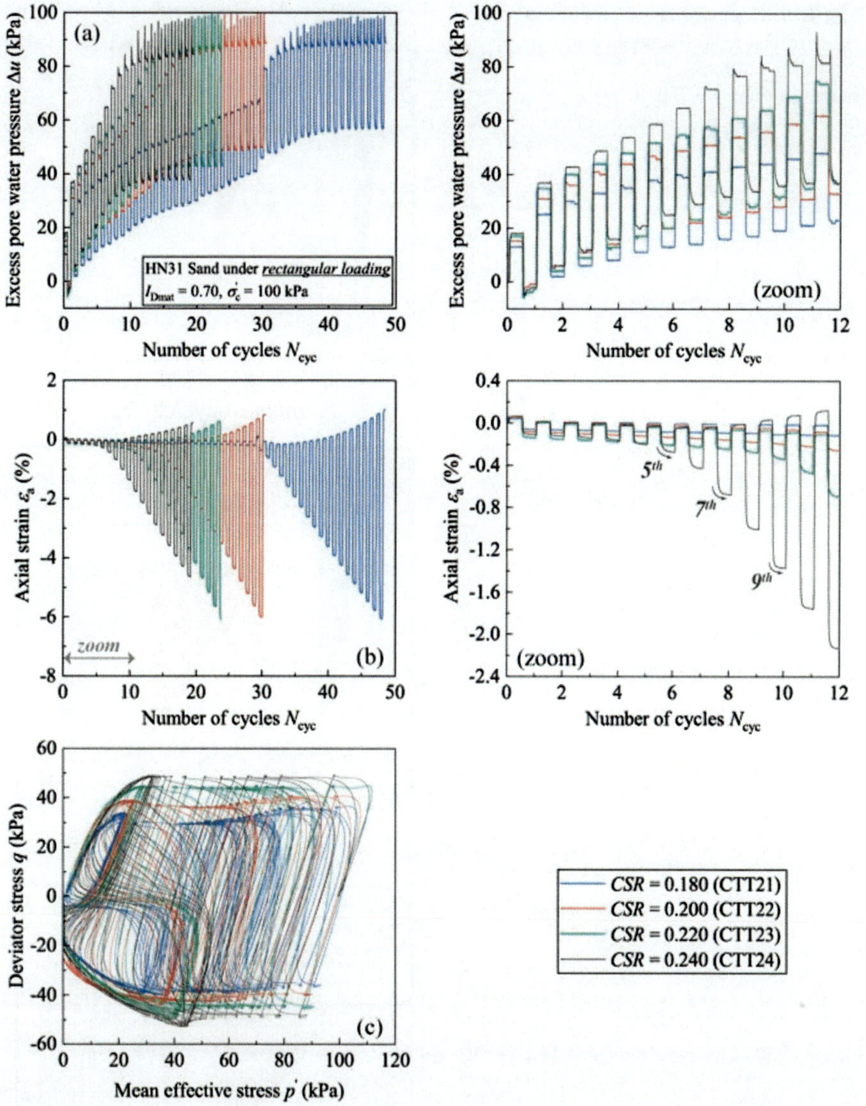

Fig. 4.25 Results obtained for HN31 sand specimens subjected to rectangular loading with $I_{\mathrm{Dmat}} = 0.70$, $\sigma'_{\mathrm{c}} = 100$ kPa: **a** excess pore water pressure-number of cycles curves; **b** axial strain versus number of cycles curve; **c** effective stress paths

a given amplitude of imposed shearing, N_{L} needed for rectangular loading therefore decreased to be smaller than the two former cases (sinusoidal and triangular). Consequently, the rectangular loading generated a more severe liquefaction concern.

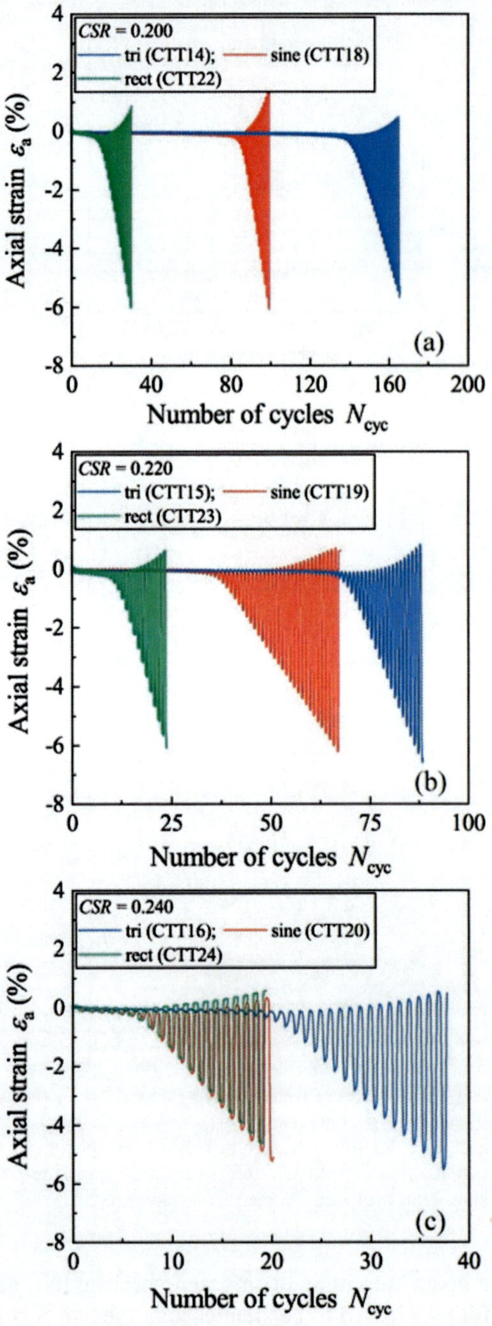

Fig. 4.26 Comparisons of experimental results between three different loading shapes: **a** $CSR = 0.200$; **b** $CSR = 0.220$; **c** $CSR = 0.240$

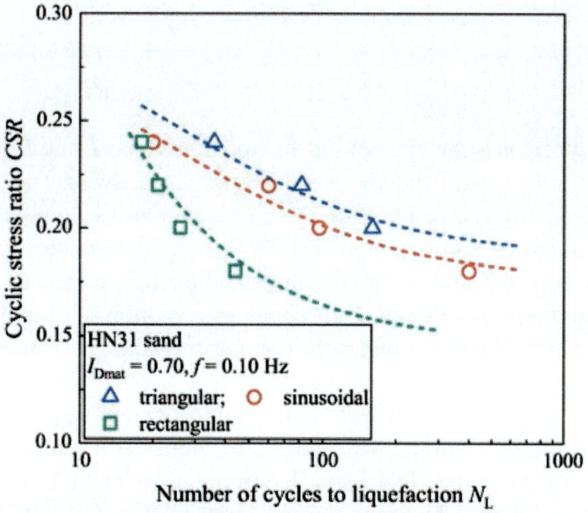

Fig. 4.27 Cyclic shear resistance curves obtained for HN31 sand for different loading shapes with $I_{\text{Dmat}} = 0.70$ and $f = 0.10$ Hz

4.6.2 Cyclic Shear Resistance Curve Versus Loading Intensity

While assessing liquefaction potential with the conventional method, N_L is usually examined in relation to *CSR* since most cyclic triaxial tests are usually performed based on sinusoidal loadings. However, the shortcoming of *CSR* is that this physical indicator only considers the impact of the amplitude value of the imposed loading and cannot express the combined influence of amplitude and loading shape. This viewpoint could explain why three different cyclic strength curves could be found in Fig. 4.27 even for test specimens with the same testing parameters. In seismic wave analysis, the Arias intensity (I_A) is an efficient measure of the strength of a given ground motion (Arias 1970; Kramer 1996; Baltay et al. 2019) and it has been proven to be a fairly suitable indicator when describing earthquake shaking to trigger seismic hazards such as landslides (Chousianitis et al. 2016) and sand liquefaction (Kayen and Mitchell 1997). It is defined as the time-integral of the square of the ground acceleration as follows:

$$I_A = \frac{\pi}{2g} \cdot \int_0^{T_d} a(t)^2 dt \tag{4.1}$$

where g is the acceleration due to gravity, and T_d denotes the duration of ground motion. While comparing the triangular/sinusoidal/rectangular loading employed in the cyclic triaxial test to a real ground motion, a new parameter termed the "loading intensity" (L_I) can then be proposed on the basis of the Arias intensity as follows:

$$L_{\mathrm{I}} = \frac{\pi}{2g \cdot m^2} \cdot \int_0^T F(t)^2 \mathrm{d}t \tag{4.2}$$

where m is the mass of the test specimen, and T and F(t) denote the period and actual force value of the imposed loading, respectively. The parameter has a dimension of velocity and can be expressed in meters per second in the SI system, like the Arias intensity. Compared with *CSR*, the new parameter has two main benefits: (i) the coupled influence of amplitude and loading shape can be systematically considered in the time-integral; and (ii) the intervention of m in the formula considers the size effect and thus, enables the comparison among test specimens having the same I_{Dmat} but different masses.

Figure 4.28 presents the evolution of the square of force value $F(\mathrm{t})^2$ for three imposed loadings with the same amplitude A but different shapes, and the surface of $F(\mathrm{t})^2$ is the algebraic value of the time-integral of the loading intensity L_{I}. It can be seen that the triangular and sinusoidal surfaces are quite close to each other, and their values are remarkably smaller than that of the rectangular one, excluding some points near the summit equal to A^2. This can logically explain why the cyclic shear resistance curve for rectangular loading is located much lower than the others.

Figure 4.29 shows the relationship between N_{L} and L_{I} for all tests examined in this section. The value of L_{I} is determined from the theoretical triangular/sinusoidal/rectangular curves by considering the fact that the imposed loading is of high quality with the adopted PID controller, and the graph demonstrates the following: (i) in the target range of imposed loading, the rectangular wave generally provides the maximum L_{I} value even with the small value of *CSR* since the influence of loading shape is fully taken into account in the formula of L_{I}; consequently, the severest liquefaction response could then be expected; (ii) following this logic, the

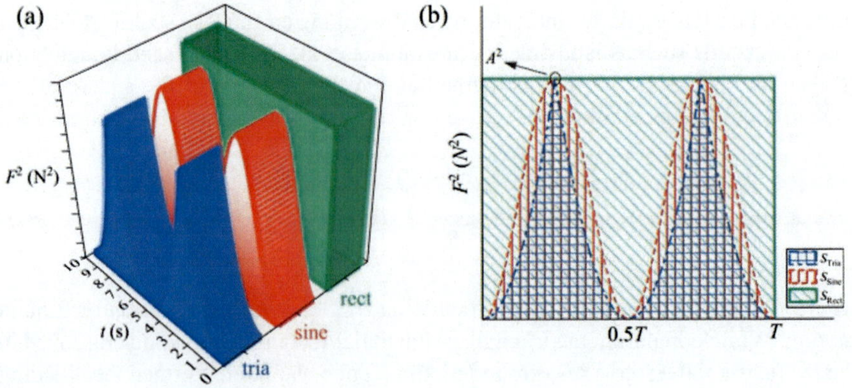

Fig. 4.28 Schematic of the square of force value against time in one cycle: **a** 3-D illustration; **b** 2-D projection

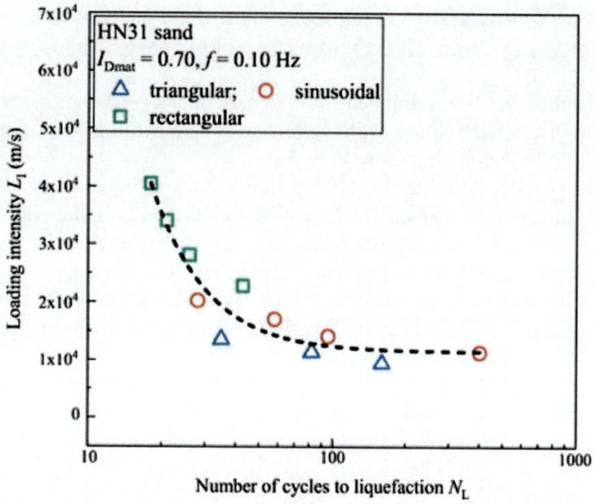

Fig. 4.29 Cyclic shear resistance curve for HN31 sand in terms of the loading intensity considering different loading shapes with $I_{\mathrm{Dmat}} = 0.70$ and $f = 0.10$ Hz

middle and moderate cases could be considered reasonable for sinusoidal and triangular shapes; (iii) for the same initial soil fabric in test specimens having the same testing parameters, *the curve falls within a narrow band and a good unification can be achieved.*

References

Arias A (1970) A measure of earthquake intensity

Baltay AS, Hanks TC, Abrahamson NA (2019) Earthquake stress drop and arias intensity. J Geophys Res: Solid Earth 124(4):3838–3852. https://doi.org/10.1029/2018JB016753

Benahmed N (2001) Comportement mécanique d'un sable sous cisaillement monotone et cyclique: application aux phénomènes de liquéfaction et de mobilité cyclique. Ph.D Thesis Ecole Nationale Des Ponts et Chaussées

Benghalia Y, Bouafia A, Canou J (2011) Comportement mécanique des sols sableux avec un indice des vides intergranulaire constant. In: 8ème Colloque National afps

Boore DM, Bommer JJ (2005) Processing of strong-motion accelerograms: Needs, options and consequences. Soil Dyn Earthq Eng 25(2):93–115. https://doi.org/10.1016/j.soildyn.2004.10.007

Chousianitis K, Del Gaudio V, Sabatakakis N, Kavoura K, Drakatos G, Bathrellos GD, Skilodimou HD (2016) Assessment of earthquake-induced landslide hazard in Greece: from arias intensity to spatial distribution of slope resistance demand. Bull Seismol Soc Am 106(1):174–188. https://doi.org/10.1785/0120150172

Dash HK, Sithara TG (2016) Effect of frequency of cyclic loading on liquefaction and dynamic preoperties of saturaed sand. Int J Geotech Eng 10(5):487–492

Feng T, Zhang L (2013) Experimental study on effect of vibration frequency on dynamic behaviors of saturatd sands. J Water Res Architect Eng 11(3):11–14, 76

Ghionna VN, Porcino D (2006) Liquefaction resistance of undisturbed and reconstituted samples of a natural coarse sand from undrained cyclic triaxial test. J Geotech Geoenviron Eng 132(2):194–202. https://doi.org/10.1061/(ASCE)1090-0241(2006)132

Guo Y, He L (2009) The influences of the vibration frequencies on liquefaction strength of saturated sands. J Disaster Prevent Mitig Eng 29(6):618–622. https://doi.org/10.13409/j.cnki.jdpme.2009.06.015

Ishihara K (1995) Soil behaviour in earthquake geotechnics. Oxford University Press

Jradi L (2018) Study of the influence of fine particles on the properties of liquefaction of sands. Ph.D Thesis University Paris-Est

Kayen RE, Mitchell JK (1997) Assessment of liquefaction potential during earthquakes barias intensity by Robert E. Kayen; Member, ASCE, and James K. Mitchell,z Honorary Member, ASCE. J Geotech Geoenviron Eng 123(December):1162–1174

Kramer SL (1996) Geotechnical Earthquake Engineering

Kumar SS, Krishna AM, Dey A (2017) Evaluation of dynamic properties of sandy soil at high cyclic strains. Soil Dyn Earthq Eng 99(May):157–167. Elsevier Ltd. https://doi.org/10.1016/j.soildyn.2017.05.016

Lee KL, Fitton JA (1969) Factors affecting thr cyclic loading strength of soil. Vibration Eff Earthq Soils Found, 71–95. https://doi.org/10.1520/STP33637S

Mulilis JP, Chan CK, Seed HB (1975) The effects of method of sample preparation on the cyclic stress-strain behavior of sands (EERC Report 75-18). University of California, Berkeley, CA, USA

Nong Z, Park S, Jeong S-W, Lee D-E (2020) Effect of cyclic loading frequency on liquefaction prediction of sand. Appl Sci 10(13):4502

Omidvar M, Iskander M, Bless S (2012) Stress-strain behavior of sand at high strain rates. Int J Impact Eng 49:192–213. Elsevier Ltd. https://doi.org/10.1016/j.ijimpeng.2012.03.004

Peacock WH, Seed HB (1968) Sand liquefaction under cyclic loading simple shear conditions. J Soil Mech Found Div 94:689–708

Polito CP (1999) The effects of non-plastic and plastic fines on the liquefaction of sandy soils. Verginia Polytechnic Institute and State University

Saxena SK, Reddy KR, Avramidis AS (1988) Liquefaction resistance of artificially cemented sand. J Geotech Eng 114(12):1395–1413

Silva WJ (1988) Soil response to earthquake gound motion

Sitharam TG, GovindaRaju L, Srinivasa Murthy BR (2004) Evaluation of liquefaction potential and dynamic properties of silty sand using cyclic triaxial testing. Geotech Test J 27(5):423–429. https://doi.org/10.1520/gtj11894

Suzuki M, Yamamoto T (2004) Liquefaction characteristic of undisturbed volcanic in cyclic triaxial test. In: 13th World Conference on Earthquake Engineering Vancouver

Tatsuoka F, Toki S, Miura S, Kato H, Okamoto M, Yamada S, Yasuda S, Tanizawa F (1986) Some factors affecting cyclic undrained triaxial strength of sand. Soils Found 26(3):96–116

Tatsuoka F, Maeda S, Fujii S, Yamada S (1983) Cyclic undrained strengths of saturated sand under random and uniform loading and their relation

Xie DY (2011) Soil dynamics. Higher Education Press, Beijing

Yoshimi Y, Oh-Oka H (1975) Influence of degree of shear stress reversal on the liquefaction potential of saturated sand. Soils Found 15(3):27–40

Zhang S, Zhang Y, Zhang L, Liu CJ (2015) Influence of confining pressure and vibration frequency on the liquefaction strength of the saturated gravel sand. J Xinjiang Agri Univ 38(1):68–71

Zhang J, Cao J, Huang S (2019) Experimental study on the effects of initial shear stress and vibration frequency on dynamic strength of saturated sands. Adv Mat Sci Eng 6:1–9. https://doi.org/10.1155/2019/3758527

Chapter 5
Shaking Table Tests

For liquefaction analysis in laboratory, monotonic and cyclic triaxial tests are widely adopted as a meaningful approach to investigating the liquefaction potential. These tests serve as the incubator to bring up the modern comprehension of sand liquefaction. Although triaxial tests already provide valuable insight into the mechanism of sand liquefaction, the understanding based on these tests is only for elementary size, hence far from representing natural soil conditions. For this purpose, large-scale shaking table tests are always preferred to better understand and predict liquefaction in a realistic engineering size. This chapter consists of three sections. The first section summarizes the experimental devices and results well-documented in the published literature and later presents the experimental setup developed in this thesis with several preliminary tests. The second section presents an advanced elastoplastic model for sand liquefaction by briefly introducing its key ingredients. An automatic calibration method based on artificial intelligence is given and the effectiveness of the proposed method is carefully evaluated with HN31 sand and Ottawa Sand F65. A numerical simulation of shaking table test is later provided, emphasising the potential boundary effects. Finally, the third section presents the ongoing development of a large-sized flexible soil container for future research.

5.1 Shaking Table Tests

5.1.1 Bibliographic Review

In the context of earthquake simulation in laboratory research, shaking table tests play an increasingly important role and are widely accepted as a valuable tool for improving the understanding of the sand liquefaction phenomenon. In such tests, the model specimens are reconstituted within finite boundaries provided by a model container. The effectiveness of the shaking table tests depends largely on the proper simulation of boundary conditions. A poorly chosen artificial boundary condition

Z. Zhu, *Influence of Fine Particles on the Liquefaction Properties of a Reference Sand*, Springer Theses, https://doi.org/10.1007/978-3-031-24299-1_5

provided by the model container may primarily change the stress and strain fields that are not present in the real field, in which the soil mass is viewed as being in a semi-infinite half space. In an ideal situation, the side walls of the model container should (i) behave exactly as the in situ soil mass, maintaining the stress field as closely as possible as the one existing in the real field; and (ii) be further capable of altering its mechanical performance so as to adapt the dynamic properties of the model specimen during base shaking. Since the shaking table test for sand liquefaction involves significant soil stiffness and shear modulus degradations with the evolution of excess pore water pressure and shear strain, both the above requirements are almost impossible regarding model container designs. Figure 5.1 demonstrates the dissimilarity of the stress/strain fields between the boundary in (a) a semi-infinite half space and (b) within a rigid model container. Provided that the side walls are much stronger than the model specimen, it is to be borne in mind that: (i) stress dissimilarity certainly occurs between elements **A** and **B** in the graph due to the lack of complementary shear stress. In addition, P-waves propagating in the horizontal direction are inevitably produced, among other superfluous wave reflections, from the side walls to the model specimen during base shaking; (ii) within the rigid model container, the deformation of soil mass near the side walls is entirely restricted by the deflection of the side walls. Thus, it is even not necessary to distinguish the deformation of soil at different depths in these areas since they are all given by the displacement of the side walls; and this is quite different from that in the imagined prototype in the real field, leading to a significant strain dissimilarity.

Although there is no perfect technical design so far to solve the aforementioned issues, some valuable attempts have been still made, aiming to minimize these boundary effects. Some well-known examples are:

(i) Cambridge Equivalent Shear Beam (ESB) container;
(ii) Laminar Shear Box (LSB);
(iii) Extension of rigid model containers (with flexible boundaries or hinged side walls).

(i) Cambridge equivalent shear beam container

At Cambridge University, ESB containers have been extensively studied in centrifuge modelling for more than twenty years. The most novel feature is the use of intermediate rubber layers separating the lightweight dural layers. During base shaking, the dural layers (see Fig. 5.2a) tend to deform with the model specimen while maintaining an appropriate state of stress. For the sake of providing the lacking complementary shear stress near the side walls, an aluminium sheet is placed and glued with sand (underlined in Fig. 5.2b).

In order to assess the performance of the proposed ESB container, an accelerometer (ACC3466) was buried in the sand in the central part of the model specimen at a slightly lower depth than ACC1552 in order to verify whether a uniform acceleration at a specific depth across the model specimen would be achieved. The recordings of these two accelerometers (ACC3466 versus ACC1552) under strong base shaking are shown in Fig. 5.3a. It can be seen that the two readings in the time domain are

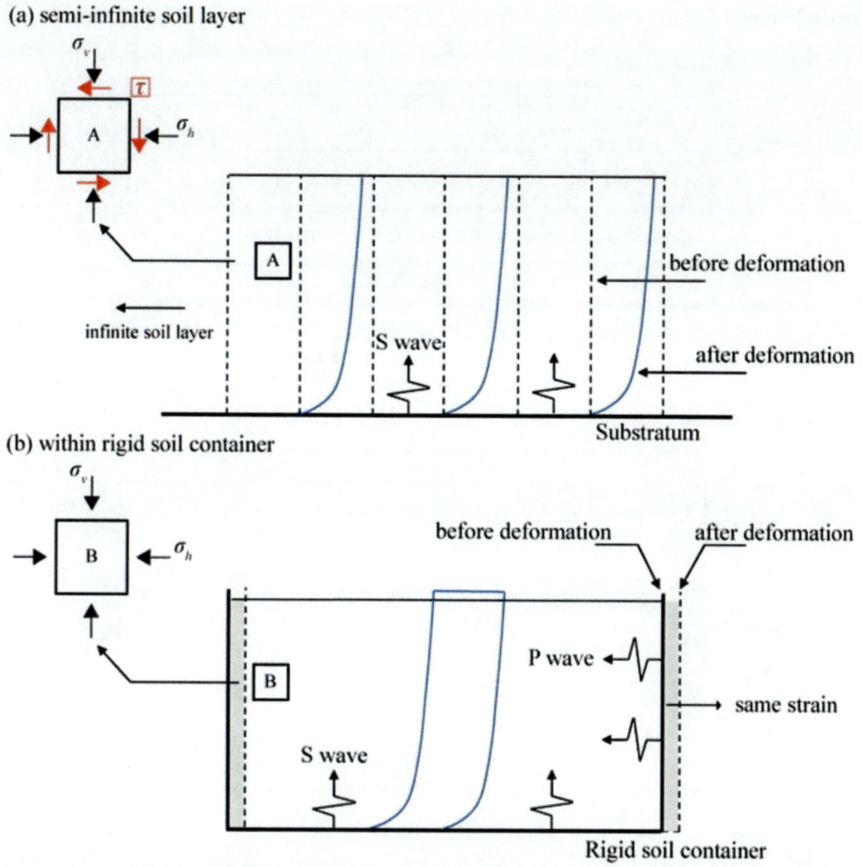

Fig. 5.1 Boundary conditions during 1D shaking: **a** in the real field; **b** within a rigid container (after Lee et al. 2012)

almost identical, except that the acceleration recorded by ACC1552 had a slightly higher amplitude (as would be expected, Zeng and Schofield 1996). In the frequency domain, the same conclusion can also be drawn (see Fig. 5.3b).

Therefore, this container design appeared to be successful and to satisfy most of the requirements for the shaking table test. However, the box is only appropriate for the use where the altering of soil stiffness is very minor. Sand liquefaction is then beyond the capabilities of this ESB design since the saturated model specimen does change its mechanical characteristics during base shaking while the stiffness of this ESB container remains constant. Recently, a deeper ESB container following the same principle was built at Cambridge University with the internal dimensions of 674 mm × 253 mm × 429 mm ($L \times B \times H$), and a detailed investigation regarding the boundary effects on dynamic centrifuge modelling (g-level = 50 g) of liquefaction in sand deposits (prepared by dry pluviation of Fraction E silica sand) has been reported in detail (Coelho et al. 2003). During the simulation of the seismic loading,

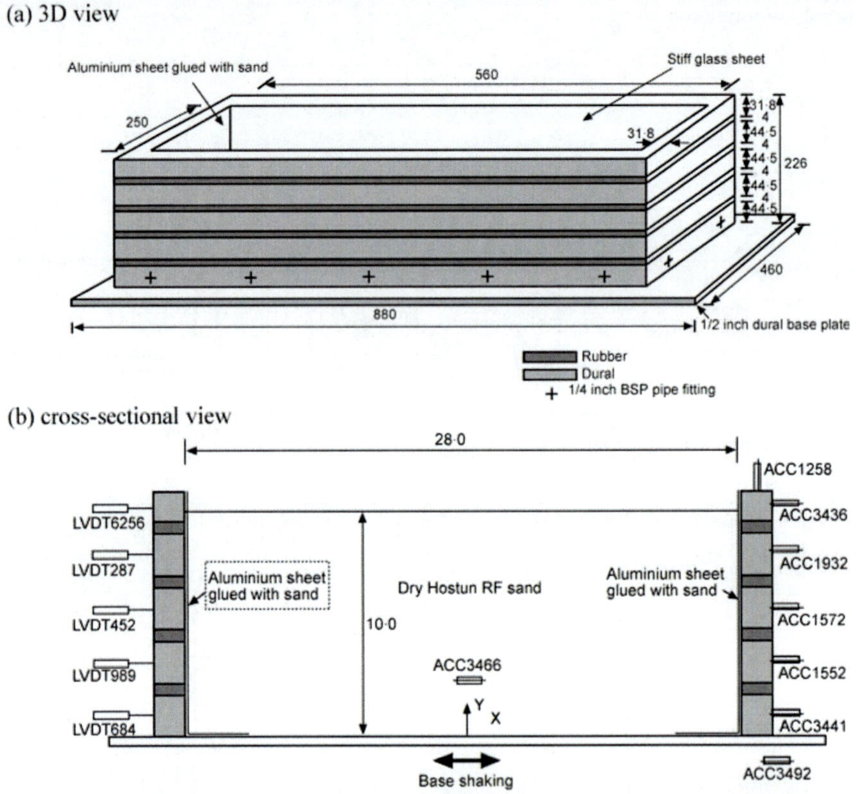

Fig. 5.2 Schematic views of Cambridge ESB container (dimensions in mm) (after Zeng and Schofield 1996)

the accelerations and pore pressures in the model specimen were carefully measured using the instrumentation layout shown in Fig. 5.4. Two model specimens were respectively built: (i) a medium-dense model specimen with a density index equal to 0.50 and (ii) a dense model specimen with a density index equal to 0.80.

Figure 5.5a presents the Fourier amplitude spectra recorded for the medium-dense model. It can be seen that (i) the FFT responses were almost identical between Position I and Position II at two different depths of 4.3 m and 12.5 m (on prototype scale); (ii) however, the discrepancy appeared to be obvious at Position III and Ring, which could be logically ascribed to the rigidity of the employed model container. Therefore, it can be deduced that for this medium-dense state, the boundary effect was well controlled for the central part of the model specimen.

Similarly, Fig. 5.5b presents the Fourier amplitude spectra recorded for the dense model specimen, at two different depths of 4.2 m and 13.1 m (on prototype scale), and none of the 4 records could be considered as being identical to the others. The above two results suggest that the boundary conditions indeed affect the dynamic soil behaviour and the ESB container is not a perfect technical solution. However, the

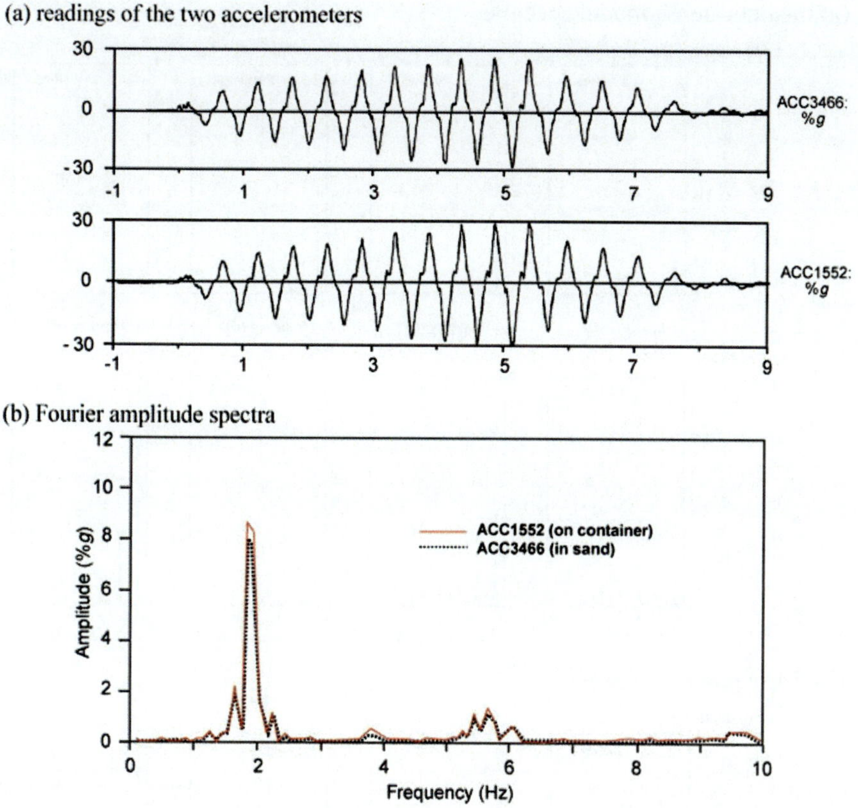

Fig. 5.3 **a** Readings of the two accelerometers; **b** the corresponding fourier amplitude spectra (after Zeng and Schofield 1996)

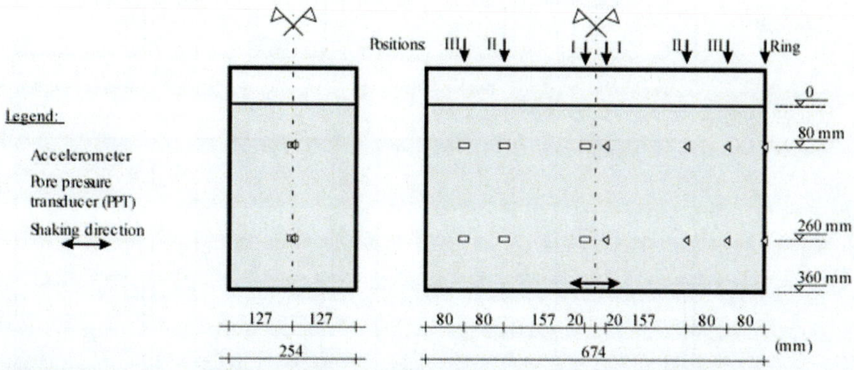

Fig. 5.4 Schematic of the sand model and instrument layout (model scale) (from Coelho et al. 2003)

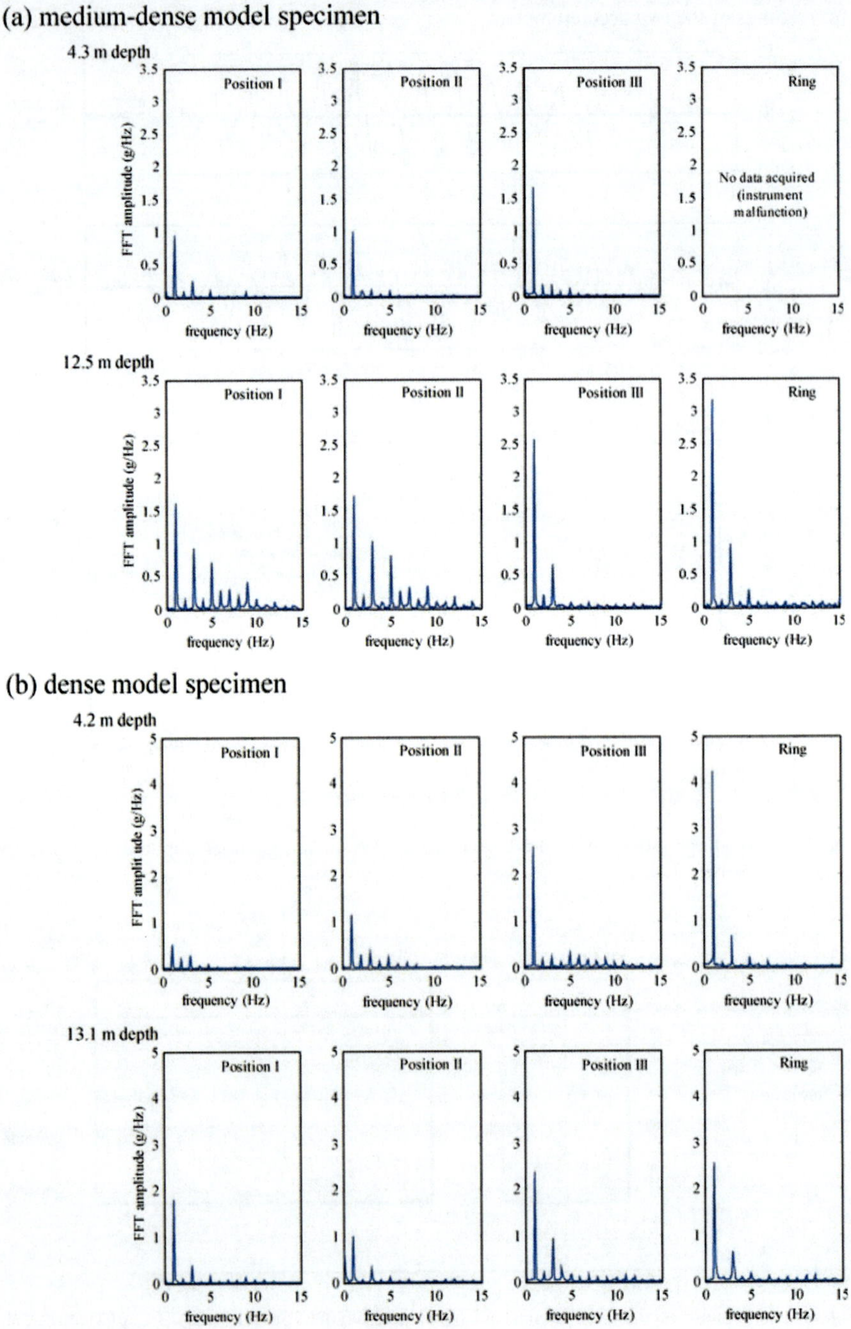

Fig. 5.5 Fourier amplitude spectra of accelerations measured in different models: **a** medium-dense sand; **b** dense sand between 25 and 35 s (prototype scale) (after Coelho et al. 2003)

placement of a loose model specimen can, to some extent, improve the performance of the ESB container.

(ii) **Laminar Shear Box**

Drawing inspiration from simple shear apparatus, the laminar shear box (LSB) is increasingly used to simulate free boundary conditions for geotechnical earthquake simulation (Suzuki et al. 2008). In general, the LSB is constructed by a series of stacked rings. The stacked rings are required to be as lightweight as possible and separated by a bearings system in order to allow a relatively free movement between adjacent layers. This design is viewed as beneficial for a liquefiable sand column since the stacked rings move in an almost frictionless manner during base shaking so that they are capable of following the rapid deformation of the model specimen and mimicking the "zero-stiffness" of a liquefied sand column. Some successful examples are well-documented in the published literature; to name a few of them:

1. A 1D laminar container (inner dimensions of 719 mm × 379 mm) constructed by the centrifuge team of Gustave Eiffel University in Nantes with the possibility of adjusting the number of stacked rings according to the height of the model specimen (Li and Escoffier 2020).
2. A 2D laminar container (inner dimensions of 1880 mm × 1880 mm) developed at NCREE (Taiwan), comprising 15 layers of frames. Each layer consists of an inner frame and an outer frame to accept the 2D ground motions (Ueng and Lee 2015).
3. A cylindrical 3D laminar container (inner diameter of 2000 mm) developed at Tianjin University, composed of 18 H-shaped laminar frames and 8 guardrails. The most extraordinary progress is the use of 144 3D limit connectors to accept the 3D ground motions (Lei et al. 2020).

– ***Gustave Eiffel University***

Figure 5.6 presents a schematic of the 1D laminar container built at Gustave Eiffel University, in which three arrays (central, middle and side) of accelerometers were buried in the model sand specimen (density index equal to about 0.50) in order to assess the boundary effect on sand liquefaction behaviour. Each array covered a range from the bottom to the free surface (from 13.5 m to 1.5 m in prototype). A series of simple sine base input signals (50 shaking events in total) with different frequencies and amplitudes was employed. After undergoing the first six shaking events of 0.1 g, the model specimen was found to be still in a loose state. The time histories of displacements recorded near the free surface ("Channel"-9 at central, "Channel"-14 at middle and "Channel"-21 on side array in Fig. 5.7a) indicated an important boundary effect since the observed amplitude values were quite different. In addition, a time lag of displacement was clearly found for the model container, as compared with the model specimen. However, with the increase in depth ($D = 7.5$ m on prototype scale), the time lag between the side walls and the model specimen remarkably decreased to be only minor. And the displacements at three positions ("Channel"-6 at central, "Channel"-12 at middle and "Channel"-19 on side array in

Fig. 5.7b) were almost identical. Therefore, the behaviour of the model specimen in the centre could be considered to be proper, mimicking a real semi-infinite soil extent subjected to 1D base shaking.

– ***National Center for Research on Earthquake Engineering (NCREE)***

Figure 5.8 presents the plan and side views of the 2D laminar container developed at NCREE in Taiwan. An in-depth comparison between model specimen responses and frame movements was reported (Chen et al. 2004) for the assessment of boundary

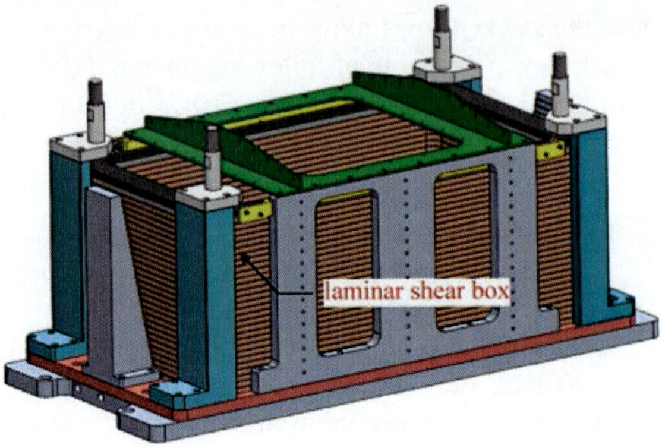

Fig. 5.6 Laminar container recently constructed at Gustave Eiffel university centrifuge laboratory (after Li and Escoffier 2020)

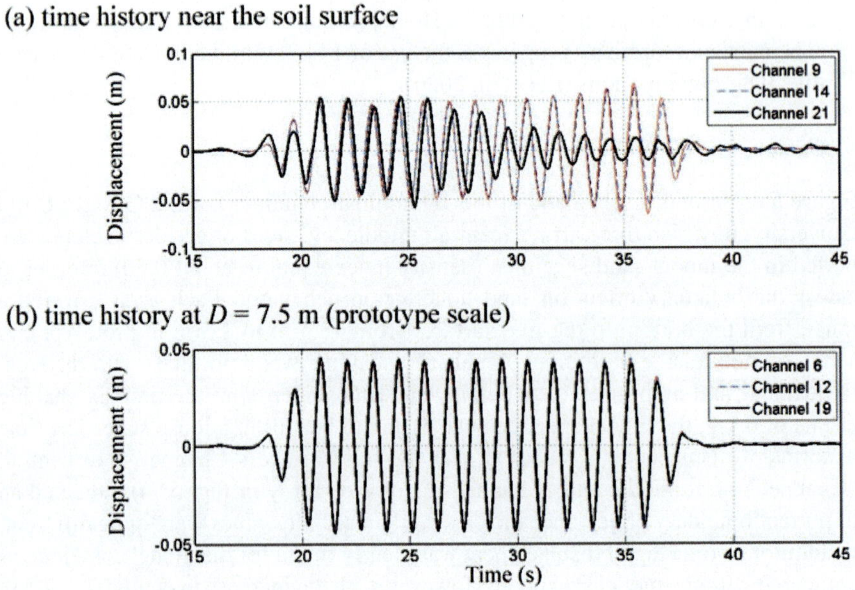

Fig. 5.7 Time histories of displacement **a** near the free surface; **b** at a depth of 7.5 m on prototype scale (after Li and Escoffier 2020)

effects. Several arrays of transducers for displacement, velocity and acceleration were installed on the frames and the outside walls. A mini-piezometer and a mini-accelerometer were also buried in the model specimen. Two shaking events with different intensities were used. Subjected to a shaking event of $A_{\max} = 0.05$ g, the time histories of accelerations and water pressure changes are shown in Fig. 5.9a. In the graph, it can be seen that liquefaction softening was far from being triggered since the generated pore pressure during base shaking remained at a low level throughout the experiment. Furthermore, the accelerations recorded on the frame and within the model specimen appeared to be fairly identical. Subjected to a shaking event of $A_{\max} = 0.075$ g, the time histories of accelerations and water pressure changes are shown in Fig. 5.9b. From the water pressure changes curve, it is found that total liquefaction softening occurred at about $t = 5$ s, and the water pressure remained at a very high level till the end of the experiment. With respect to the time histories of accelerations, two phenomena could be identified: (i) before attaining sand liquefaction ($t < 5$ s), there was no obvious difference between the inner and outer frames; (ii) whilst after the liquefaction initiation ($t > 5$ s), the acceleration of the inner frame decreased to be much smaller and the acceleration of the outer frame became somewhat irregular with the presence of spikes.

These observations could be taken as logical and explained by the sudden loss of stiffness of the model specimen due to the triggering of liquefaction softening. After sand liquefaction took place, the model specimen developed a "flow-type" response and behaved like a liquid. As a result, the frames of the model container lost their capacity to perfectly follow the transient movement of the model specimen, resulting in a significant boundary effect. However, prior to the attainment of sand liquefaction, the boundary effect was very limited; thus, the experimental data could be viewed as being of great value in highlighting the response of a soil mass in a semi-infinite half space during seismic shaking.

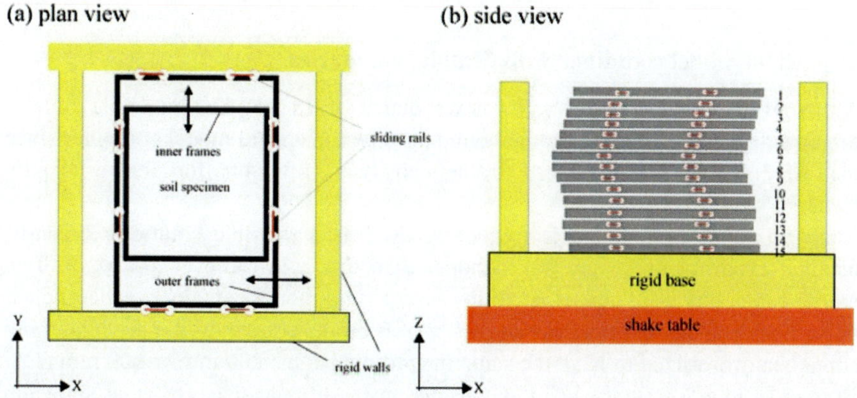

Fig. 5.8 Schematic views of biaxial laminar shear box developed at NCREE (Taiwan) (after Ueng and Lee 2015)

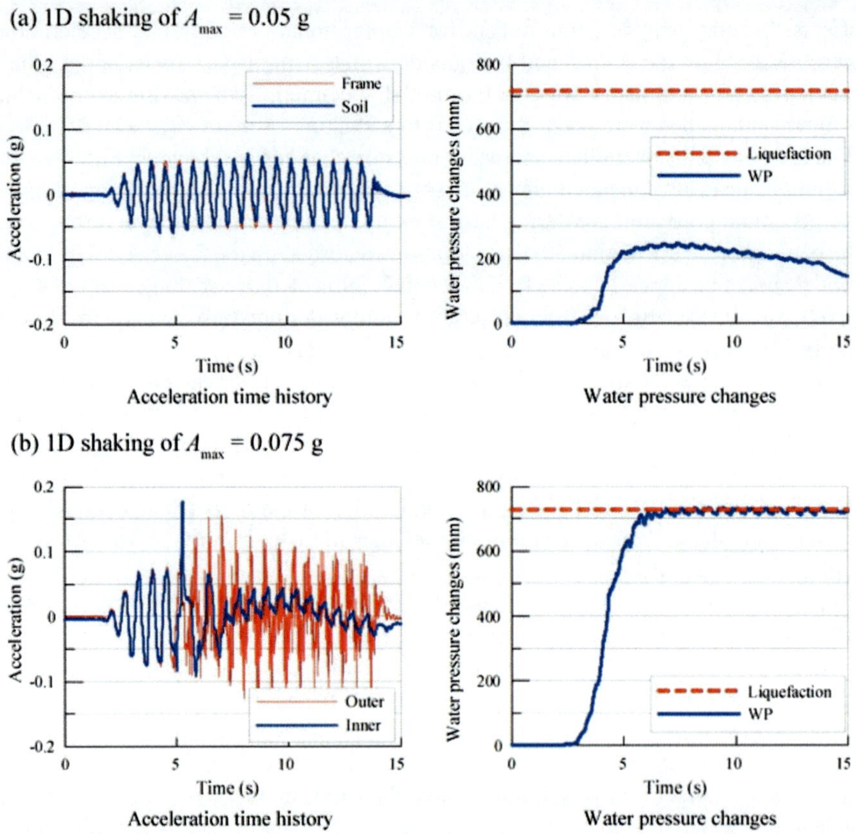

Fig. 5.9 Time histories of acceleration and water pressure changes for a shaking event: **a** A_{max} = 0.05 g; **b** A_{max} = 0.075 g (after Chen et al. 2004)

(iii) **Rigid model container with flexible boundaries**

Although the benefit of LSB is well-known, there is still a need for a workable solution in a practical sense to mitigate the boundary effect in a rigid model container since the LSB container is sometimes extremely costly and often unaffordable, especially for large-scale tests.

To this end, the sponge is frequently used as a possible solution. Recently, detailed experimental work was conducted in this area (Kim et al. 2020). Two model containers were chosen as follows: (i) a rigid model container made of clear polyethene of 10 T in thickness and (ii) a LSB model container. In particular, both containers were sized to have the same inner dimensions: 200 mm × 600 mm ($L \times B$) in order to eradicate the need to consider any additional size effects, as shown in Fig. 5.10a. During base shaking, accelerations at different positions and depths were measured using the instrumentation plans shown in Fig. 5.10b. In addition, sponges of different thicknesses (5, 10 and 15 cm) were attached to the side walls. An artificial seismic wave was used as the base excitation. For further analysis, the acceleration spectra (SA) were taken as a comparison basis to quantity the boundary effect on the dynamic issue.

All experimental outcomes are summarized in Fig. 5.11, respectively. The performance of the rigid model container alone is assessed in subplots (a)–(c), and it can be seen that the accelerations recorded at different positions and depths obviously dispersed with each other, indicating a significant boundary effect. The performance of the LSB model container is assessed in subplots (d)-(f), and it can be seen that

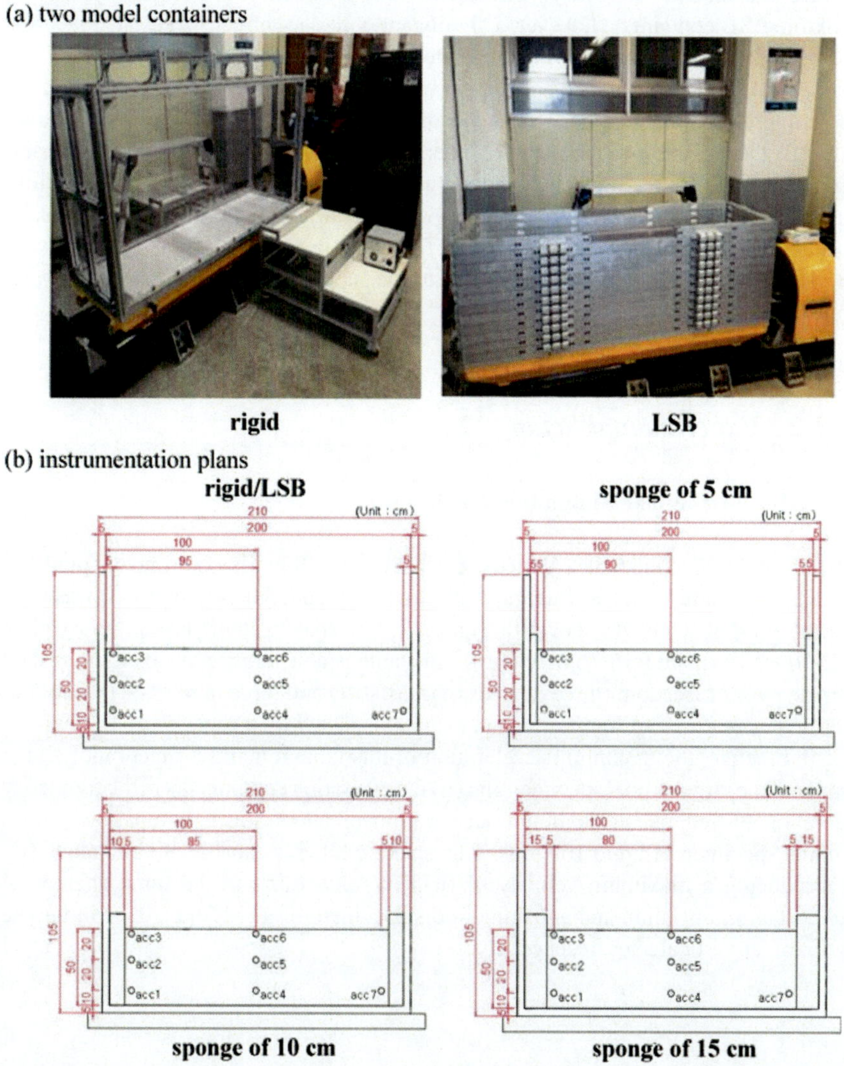

Fig. 5.10 **a** Model containers used; **b** instrumentation plans (after Kim et al. 2020)

the accelerations recorded at the bottom (acc1 and acc4) and at half-height (acc2 and acc5) were almost the same, indicating that the model specimen in these areas experienced a proper simple shear condition during base shaking, highly similar to the free field behaviour of a semi-infinite half space. However, near the free surface (acc3 and acc6), two acceleration readings were not consistent. This finding is in good agreement with the aforementioned experimental results (Li and Escoffier 2020). The performance of sponges of various thicknesses is assessed in the subplots (g)–(o), and it can be seen that the coupled acceleration spectra at each depth of the rigid model container with the attached sponges showed almost the same tendency in the centre and the side wall (acc1 versus acc4 and acc2 versus acc5), highly like the flexible LSB container. However, a significant boundary effect still existed near the free surface. While attaching flexible materials such as sponges to the side walls of a rigid model container, it is to be borne in mind that (i) the boundary effect would not be completely mitigated near the free surface, similar to the LSB model container and (ii) the selection of the flexible materials is prevailing since different materials contribute to different impacts of reducing boundary conditions. Thus, some preliminary tests should be carefully designed with the target model specimen to make a suitable compromising choice, especially for dynamic liquefaction analysis involving important soil stiffness degradation due to the rapid build-up of excess pore water pressure.

5.1.2 *Experimental Setup*

5.1.2.1 Earthquake Simulation Facility

A large-scale shaking table (Azalée) developed at the Tamaris experimental platform of CEA (Commissariat à l'Énergie Atomique et aux énergies alternatives) was used in the present work. As shown in Fig. 5.12, this shaking table has an upper plate (25 tons) in square form (6 m × 6 m). This table is able to generate excitations with six degrees of freedom (three degrees in translation and three in rotation) using four vertical actuators and four horizontal actuators, each with a capacity of 1000 kN.

In practice, the maximum acceleration of this table remains a function of (i) the mass of the target model specimen and (ii) the excitation components in the frequency domain. In general, a maximum acceleration of about 1 g is compatible with a model specimen of up to 100 tons. This shaking table is capable of simultaneously reproducing a maximum velocity of 0.35 m/s according to the three degrees of freedom in translation and a maximum displacement of ± 12.5/ ± 10 cm along the horizontal/vertical axes, respectively.

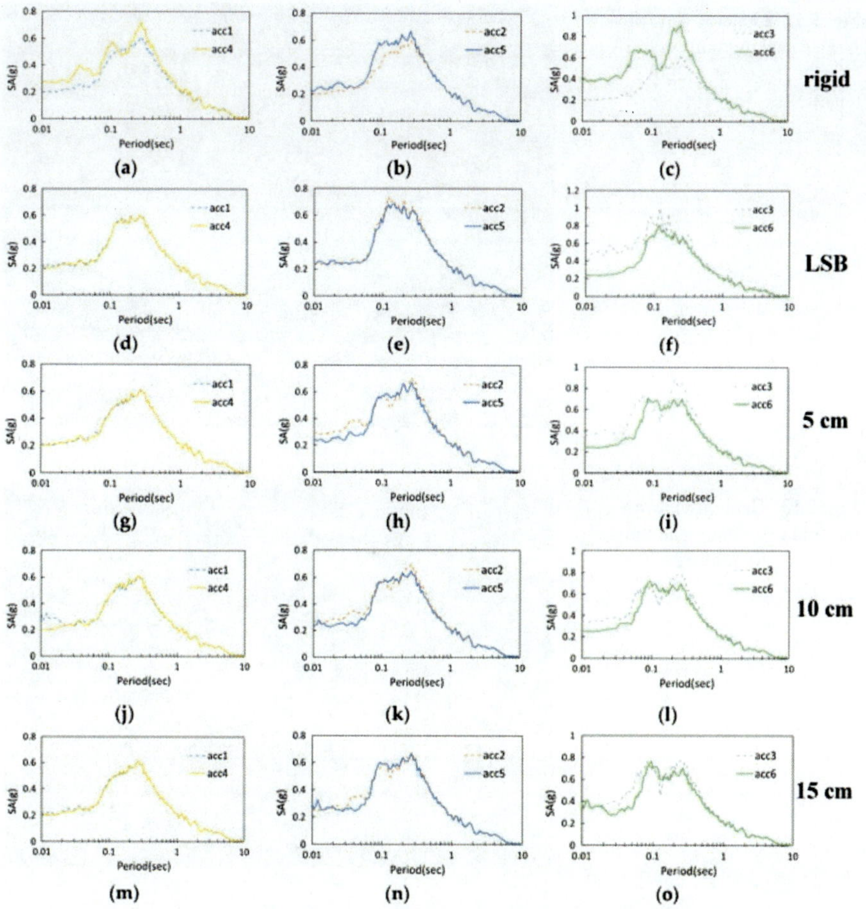

Fig. 5.11 Acceleration spectra by depth for **a–c** rigid model container; **d–f** LSB model container; **g–i** sponge of 5 cm; **j–l** sponge of 10 cm; **m–o** sponge of 15 cm (after Kim et al. 2020)

5.1.2.2 Model Soil Containers

As mentioned in the literature review, the specimen preparation method (§1.3.1) has been repeatedly proved to be a prevailing factor governing the sand liquefaction response. With this in mind, two soil containers have been designed and fabricated in the present work, respectively. In order to highlight the possible impact of the experimental procedure, the first one has a small size with inner dimensions of 0.87 m × 0.87 m × 1.20 m ($L \times W \times H$). The objective was only to evaluate the different sand preparation and saturation methods in terms of homogeneity throughout the model specimen instead of cautiously addressing the liquefaction issues. On the contrary, the second one is the final soil container with dimensions of 4 m × 4 m

Fig. 5.12 General view of Azalée shaking table at CEA (Azalee.jpg (400 × 300) (cea.fr))

Fig. 5.13 General view of the small container and the installed geomembrane

× 2 m ($L \times W \times H$), with the purpose of exploring the possible sand liquefaction during base shaking.

Regarding the small container, a waterproof geomembrane bag is initially placed and fixed on the side walls prior to placing the sand. With the presence of the geomembrane, the box can house a net specimen of 0.87 m × 0.87 m × 1 m ($L \times W \times H$). The bag is capable of preventing soil and water from seeping out of the box, especially during the saturation stage. It should be mentioned that the selected bag is deemed to be sufficiently flexible in order not to govern the sand liquefaction response (Fig. 5.13).

5.1.2.3 Preliminary Work Concerning Reconstitution Method

In general, air pluviation is a common technique for preparing sand specimens for large-scale shaking table tests. The deposition intensity (or pluviation flow rate) and drop height are two important parameters that need to be accurately calibrated in preliminary tests in order to gain a high-quality homogenous test specimen (Rad and Tumay 1987). However, this procedure could not be, with ease, realised for the target specimen size (4 m × 4 m × 2 m) in the present work since the freefall of sand grains under gravity during the procedure would probably cause thick sand dust in the testing hall. Moreover, the manufacturing of a sand hopper able to house such a voluminous granular assembly of more than 32 m^3 and the accurate controlling of its upward movement to place sand grains, are almost unachievable. Thus, this method is not covered by the present work. As an alternative, the dry tamping method (DT) and the white noise (WN) vibration method were investigated.

In order to gain a complete insight into the internal homogeneity of the sand specimen, the vane shear test was taken as a yardstick to recognize the quality of each reconstitution method. The vane shear test is an in situ geotechnical measuring method to estimate the soil strength without introducing significant disturbances. The test is relatively quick and provides a cost-effective manner to obtain the field shear strength in relation to its relative density; therefore, it is widely adopted in geotechnical investigations. The hand-held vane shear device manufactured by YING AN YANG (in Su Zhou, China) was employed. The entire system (as shown in Fig. 5.14) involves: (i) a torsiometer with a precision greater than 10%, (ii) six extension bars with a maximum measuring depth of up to 3 m and (iii) four different sizes of vane heads according to different soil rigidities, as tabulated in Table 5.1.

In order to evaluate which kind of vane shear measurement might be expected if the homogeneity or the relative density is well controlled throughout the test specimen, an elementary cylindrical specimen (18.7 cm in diameter and 40 cm in height, as shown in Fig. 5.15a) was reconstituted with 5 layers by adopting the DT method.

In each layer, the mass of sand was calculated through the density index of sand matrix I_{Dmat} equal to 0.50. Afterwards, sand of each layer was carefully introduced into a membrane-lined split mould followed by slight compaction using a hand-held tamper. The vane shear readings were initiated at a depth of $D = 10$ cm and ended

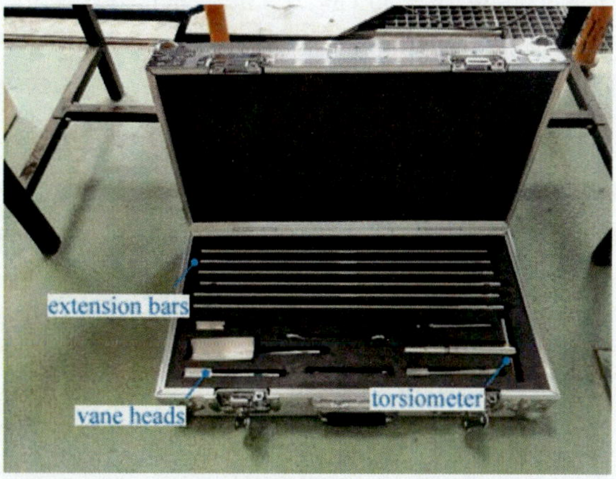

Fig. 5.14 General view of the hand vane shear device

Table 5.1 Parameters of different vane heads

Reference	Size (mm)	Measuring range (kPa)	Coefficient
A (Standard)	20 × 40	0–130	1
B	16 × 32	0–260	2
C	25.4 × 50.8	0–65	0.5
D	50.8 × 101.6	0–8.125	0.0625

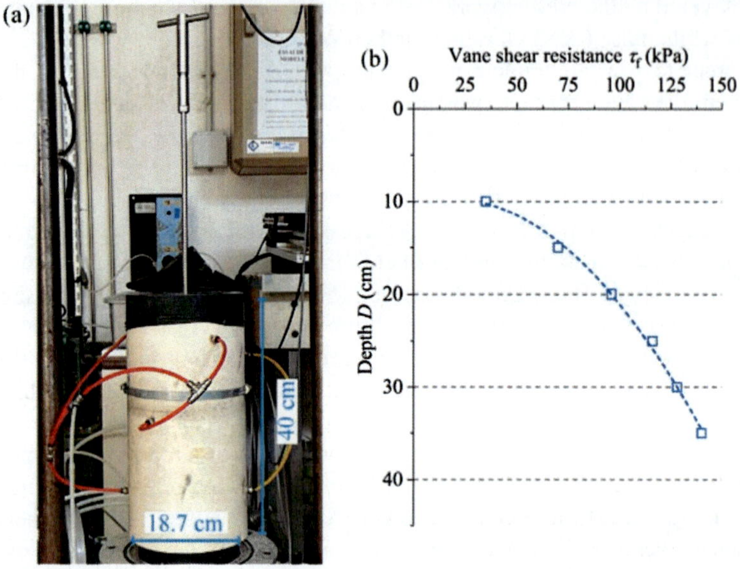

Fig. 5.15 **a** View of the vane shear test on an elementary sand specimen; **b** vane shear test measurements

at a depth of $D = 35$ cm. The measurements were taken in the central part of the test specimen and the reading frequency was set to be every 5 cm. It appears from Fig. 5.15b that the relationship between the vane shear resistance τ_f and the depth D was almost linear, although a slight parabolic tendency was indeed found between 10 and 20 cm. This observation could be taken as reasonable by the fact that as granular material, a sand specimen generally has a greater effective friction angle under a lower consolidation stress standing for a shallow depth. However, this minor discrepancy does not influence the conclusion that could be drawn from this trial that the soil shear strength such as vane shear measurement should increase quasi-linearly with the increment in depth if a soil column is relatively homogeneous (Turan et al. 2009).

A model sand column of about 60 cm (initial height) was reconstituted in the small box by adopting the DT method through two successive layers to attain a target density index of sand matrix I_{Dmat} equal to 0.70 (corresponding to about 50 cm in final height). A solid wooden plate was carefully put on the surface of each layer and a minor distance between the wooden plate and the side wall was maintained during the compaction in order not to destroy the fragile geomembrane. A heavy hammer (see Fig. 5.16a) was lifted and freely thrown down to hit the wooden plate; accordingly, the wooden plate received the freefall energy so as to densify the model sand column below. This procedure was repeated several times at different positions of the model sand column by altering the placement of the wooden plate. The vane shear readings were then performed at 5 different positions (the central part and 4 positions close to the 4 corners) and at 4 different depths: 10 cm, 20 cm, 30 cm and 40 cm, respectively. Note that the standard vane head A was employed for this dense state. It was experimentally found that the vane shear readings at the same depth but at different locations fell within a narrow range since the experimental readings dispersed within $\pm$ 15% of the mean value. Consequently, all data are averaged in Fig. 5.16b for better viewing. In the graph, it can be seen that the calculated mean value could be satisfactorily fitted by a single straight line, revealing that the DT method is capable of generating a relatively homogeneous model sand column for the shaking table test. However, although the final settlement after the compaction could not be precisely measured in such a sized container, the author acknowledges that the final value was certainly an underestimation to attain $I_{\mathrm{Dmat}} = 0.70$. This argument signifies that (i) as for the large-sized shaking table test, the manual dry tamping procedure should be discouraged, especially for a model sand column in a very dense state; (ii) mechanical tools such as dynamic compactor might be used as an alternative to replacing the exhausting manual tamping procedure.

After the DT method, the same model sand column was also subjected to a white noise (WN) base excitation, using a smaller shaking table named "Vésuve" (3.8 m ×

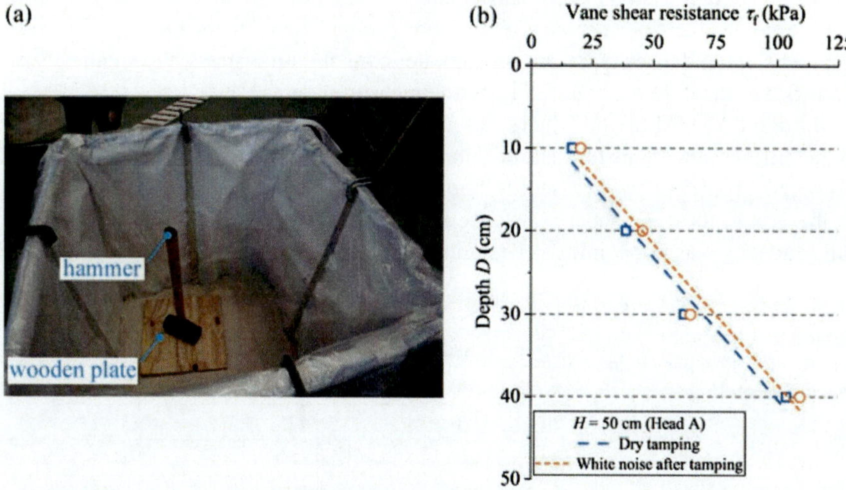

Fig. 5.16 **a** View of the dry tamping method in the small soil container; **b** vane shear measurements by adopting the dry tamping and white noise method

Fig. 5.17 General view of Vésuve shaking table at CEA (Vesuve.jpg (400 × 274) (cea.fr))

3.1 m on each side), as shown in Fig. 5.17. This shaking table can generate mono-axial excitations and withstand a total mass of up to 20 tons. The white noise was terminated when reaching a final settlement corresponding to $I_{\mathrm{Dmat}} = 0.66$. Vane shear measurements were immediately conducted after the vibration. Two phenomena could be identified in Fig. 5.16b: (i) the subsequent WN could still increase the soil density, justified by the increment in vane shear readings as compared with the previous one; (ii) the subsequent vibration generally maintained a good "quasi-linear" relationship between τ_{f} and D, pointing to a good internal soil homogeneity.

To completely examine the effectiveness of the WN method, a new model sand column was built with 5 layers. The WN shaking events were initiated after each layer had been placed at the elevations tabulated in Table 5.2. It should be mentioned that the first two layers were initially much looser than expected; thus, more shaking events were then used so as to densify the soil column completely. The vane shear measurements were performed after the above stage. Owing to the fact that the model sand was much deeper representing a greater consolidation stress, the smallest vane head B was taken in order to facilitate the measurement.

Similar to the DT method, there was no significant variation in vane shear readings at the different locations for the same elevation; therefore, all experimental data were averaged. Figure 5.18 presents the relationship between τ_{f} and D for this sand column. In the graph, the vane shear resistance initially increased with the depth; however, this tendency was not continually maintained after $D = 50$ cm and a decrease in the

Table 5.2 Elevation information of each layer and shaking events

Layer	Initial height (mm)	Final height (mm)	Shaking event
1	240	230	3
2	450	425	3
3	675	650	1
4	880	N/A	2
5	1100	1050	2

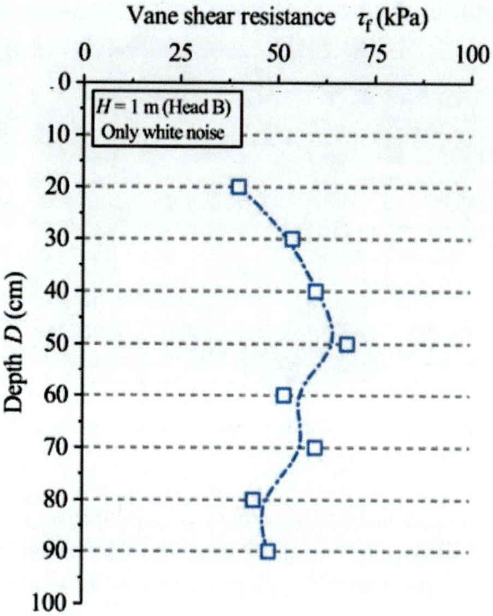

Fig. 5.18 Relationship between the vane shear resistance and the depth by adopting the pure white noise method

vane shear resistance was clearly observed. This phenomenon suggests that the use of the WN method only, in general, led to a non-uniform distribution.

Finally, we sum up here the advantages and shortcomings of each reconstitution method. The DT method generally provides a good distribution of sand, but the tamping procedure is very time-consuming and, sometimes, beyond the reach of human power, especially for a very dense sand column. In addition, the WN method could be added following the DT method to achieve a higher density state without remarkably altering the homogeneity of the model test specimen. However, the WN method cannot be used alone even if it is normally easier and more rapid. This defect might be attributed to the propagation of shear waves into a finite space with the presence of a rigid soil container: shear waves are inevitably reflected from the side walls to the model sand specimen in a non-uniform manner along the side walls. Thus, as for liquefaction-related issues, the DT method could be viewed as beneficial for building a homogenous model sand column in the medium-dense state. Besides, a mechanical tool such as a dynamic compactor can be employed in order to overcome its inherent shortcoming.

5.1.2.4 Preliminary Work Concerning Saturation Method

With the purpose of examining the feasibility of the specimen saturation method, a preliminary test was carried out using the small container. Provided that the small container has no bottom part permitting the reception of carbon dioxide and water, an inlet tube was first installed in the geomembrane and a drainage layer of about

(a) inlet tube (b) drainage layer

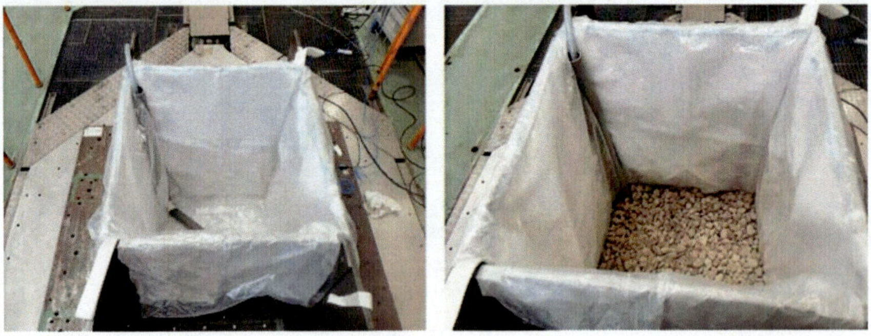

Fig. 5.19 Saturation process: **a** view of the inlet tube; **b** view of the drainage layer

20 cm was later deposited which later served as a rigid "macro porous stone" in order to diffuse the air and liquid in a static and uniform manner (Fig. 5.19).

After carefully placing a thin geotextile, a model specimen (80 cm) of $I_{Dmat} = 0.50$ (or void ratio $e = 0.83$) was built using the DT method as previously presented. A pore water transducer was installed at the bottom of the sand specimen to record the possible variation of pore water pressure during saturation. Afterwards, the container was totally enclosed with an air-tight cover, as shown in Fig. 5.20. Carbon dioxide (CO_2) was firstly flushed through the dry test specimen and the inlet flow was controlled as small as possible in order not to destroy the brittle soil fabric. Due to the fact that CO_2 is much heavier than air, the air bubble was gradually evacuated through the outlet line from bottom to top. To ensure safety, an aeration device was activated and kept working during the injection. Finally, the complete fill of CO_2 gas was carefully checked by a sensible CO_2 detector: *the colour of the device became red if the CO_2 concentration at the top part of the container was greater than 2000 PPM.* Using the same inlet line, the normal (not de-aired) water was then injected. In like fashion, the inlet flow of pore water was well controlled being only 2 L per minute using a fluxmeter (as shown in Fig. 5.20). Figure 5.21 displays the relationship between the pore water pressure (at the bottom) and the water level (manual recording roughly every 10 min). As for the entry water, it took approximately 1 h to pass through the drainage layer, during which the pore water pressure remained null.

After this, an almost linear tendency was recorded by both the pore water transducer and manual recordings. This linear tendency implies that this model specimen had been successfully built in a uniform manner since the inlet flux of water was controlled to be constant, irrespective of the evolution of waterhead. This finding indeed reconfirms the conclusion drawn from the previous section; namely, the dry tamping method is a consistent method to meet the homogeneity requirement. After about 3 h, the pore water pressure arrived at a value of about 8 kPa, corresponding to a saturated sand column of about 80 cm. Finally, the moisture content (the mass of pore water divided by the *total surrounding mass*) was carefully measured (see

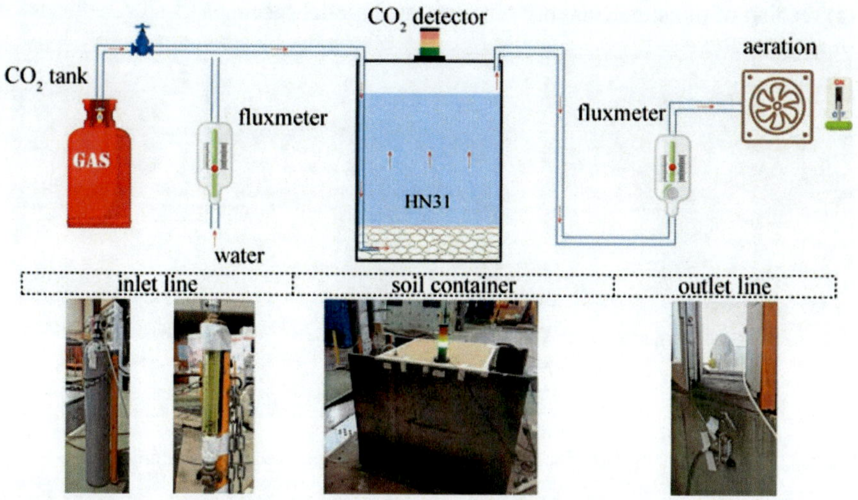

Fig. 5.20 Schematic and photographic view of the injection of CO_2 and water

Fig. 5.21 Relationship between pore water pressure (smoothed by the adjacent-averaging method) and water level

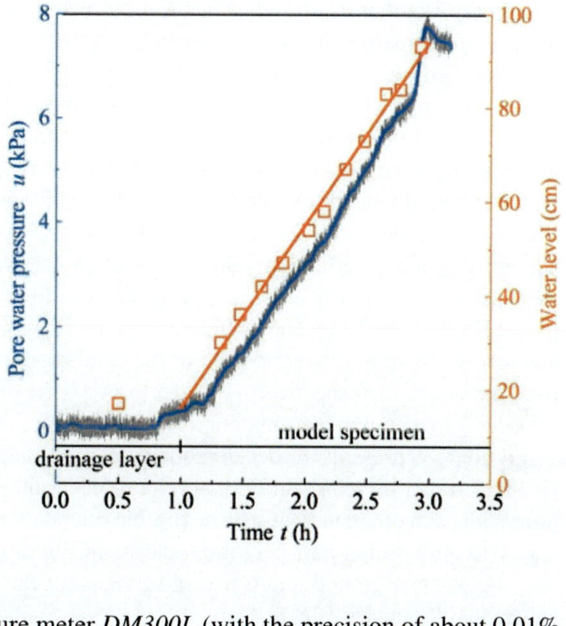

Fig. 5.22a) by a digital moisture meter *DM300L* (with the precision of about 0.01% in double precision mode, manufactured by Tuoke in China), allowing the saturation degree S_r to be subsequently determined using the given void ratio e equal to 0.83. For this trial, the obtained S_r was found to be very close to 100%.

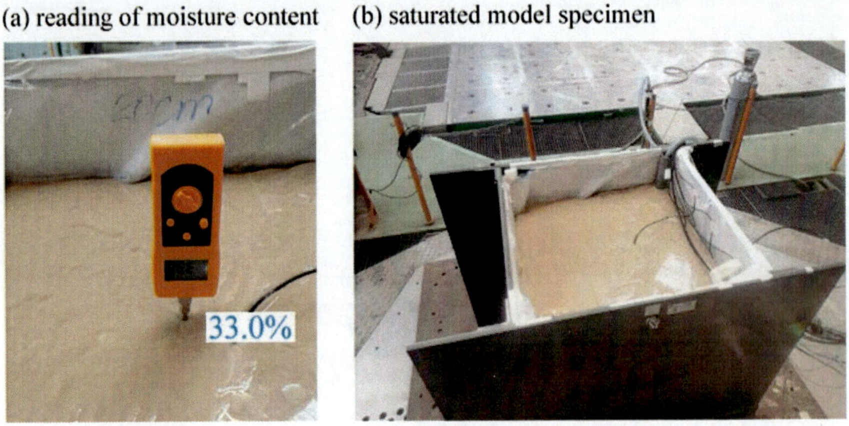

Fig. 5.22 Photographic views of **a** moisture content reading; **b** specimen in saturated state

5.1.2.5 Preliminary Work Concerning Sand Liquefaction

Due to the fact that it is too expensive and time-consuming to assess the reliability of the entire experimental setup directly using the large soil container, a preliminary test was then performed with the small one. The objective was to verify whether the sensors (especially pore water transducers with a precision greater than 1% in Fig. 5.23b) could work properly and roughly capture the typical sand liquefaction response in such a test. A sand specimen of 80 cm was prepared with four layers using the DT method to achieve a density index I_{Dmat} of 0.50 or $e = 0.83$ (see Fig. 5.23a) and later saturated with the method described previously. After that, the moisture content was measured and a value of about 34.4% was obtained, conforming to a good saturation (S_r very close to 100%). For simplicity, a modified sinusoidal wave ($f = 2$ Hz, $a_{\max} = 0.75$ g and a total duration of 25 s) was used as the base excitation. Two arrays of sensors were buried in the sand model (at the central part and 5 cm from the side wall, respectively) in order to assess the possible boundary effect during shaking. The detailed instrumentation layout is provided in Fig. 5.24.

Figure 5.25 presents the accelerations recorded at different depths during the base shaking. It can be seen that both accelerations at the bottom were almost overlapping with each other, indicating a negligible boundary effect. The same phenomenon was also observed at half-height, as shown in Fig. 5.25b. However, a vast discrepancy between *sensors* 5 and 6 was observed near the top free surface, indicating a substantial boundary effect.

Figure 5.26 presents the excess pore water pressures recorded at different depths during the base shaking. The dotted lines in all subplots stand for the triggering of sand liquefaction ($\Delta u = \sigma'_v$). For the bottom, it can be seen that Δu steadily increased as the base shaking was initiated. However, the development of Δu remained too low to trigger the sand liquefaction. After the end of base shaking ($t = 25$ s in Fig. 5.26a), Δu was quickly dissipated in the following 25 s. As compared with

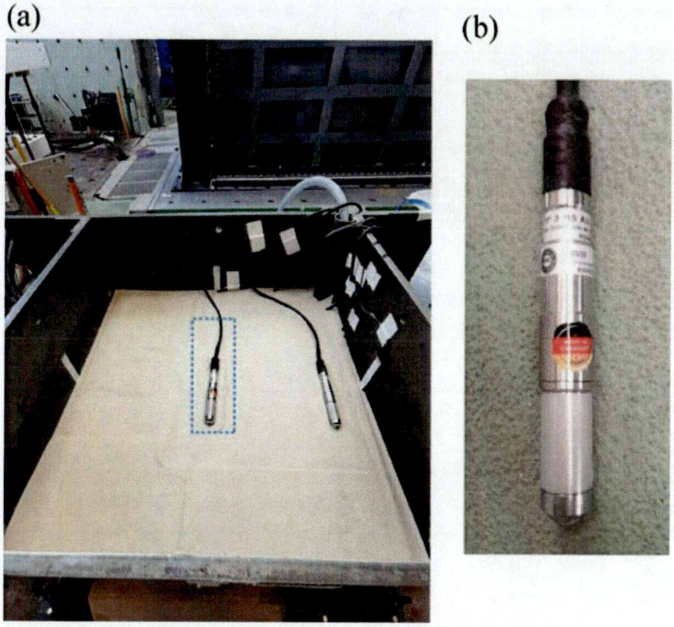

Fig. 5.23 **a** General view of the sand model; **b** pore water pressure transducer

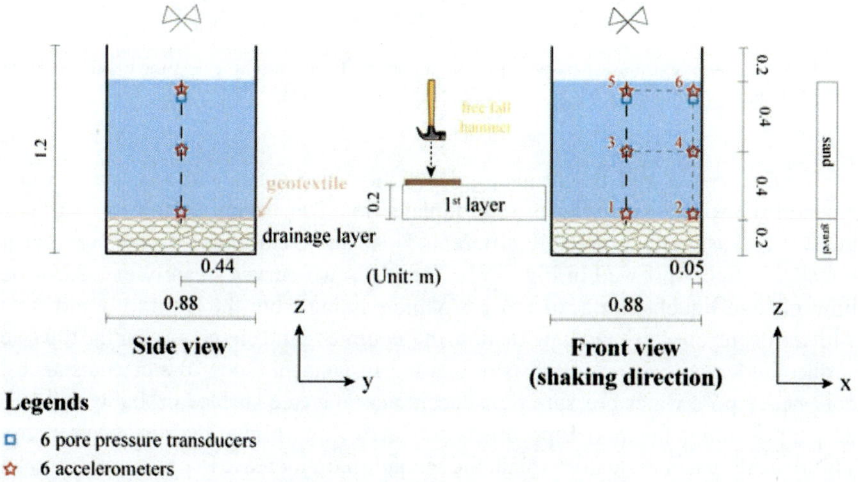

Fig. 5.24 Schematic of the sand model and instrument layout

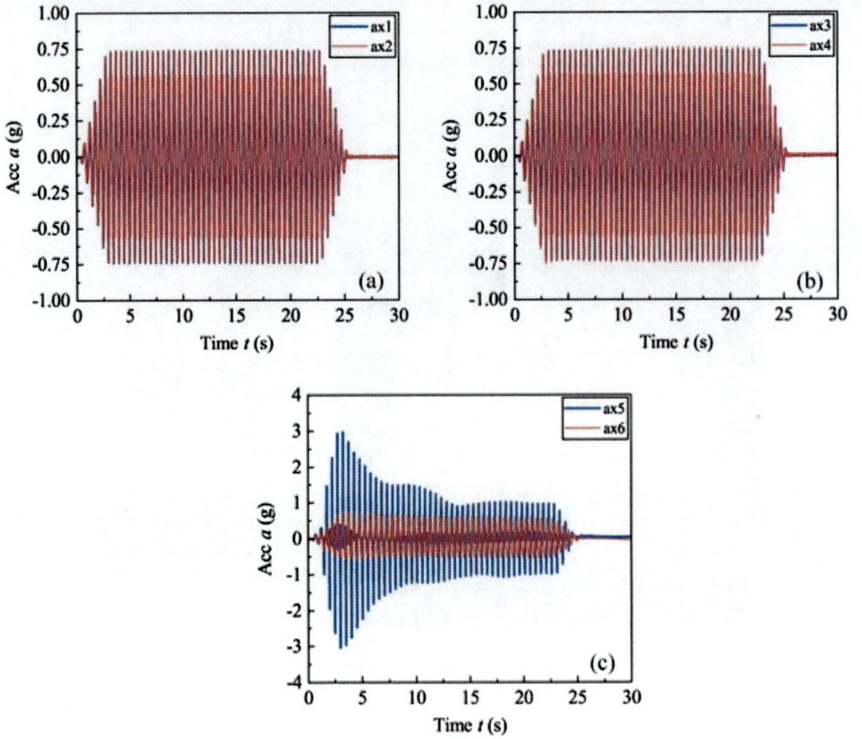

Fig. 5.25 Accelerations during base shaking at: **a** bottom; **b** half-height; **c** near the top free surface (after FFT bandpass filter with $f_{cut1} = 0.10$ Hz and $f_{cut2} = 20$ Hz)

the bottom, Δu at half-height also quickly increased and could arrive at a value much closer to the triggering of sand liquefaction. Regarding the top free surface, the two curves were completely different. This can be explained by the significant boundary effect observed in Fig. 5.25c. Besides, two curves steadily exceeded the limit of sand liquefaction ($\Delta u > \sigma'_v$). While passing from the solid to liquid state with the occurrence of sand liquefaction, the saturated sand elements near the top free surface underwent a transient loss of bearing resistance. Under the circumstances, both heavy pore water pressure transducers near the free surface probably fell to a deeper position in the model specimen associated with a higher hydrostatic pressure. This reasoning can logically explain the second sharp increase in pressure, as shown in Fig. 5.26c.

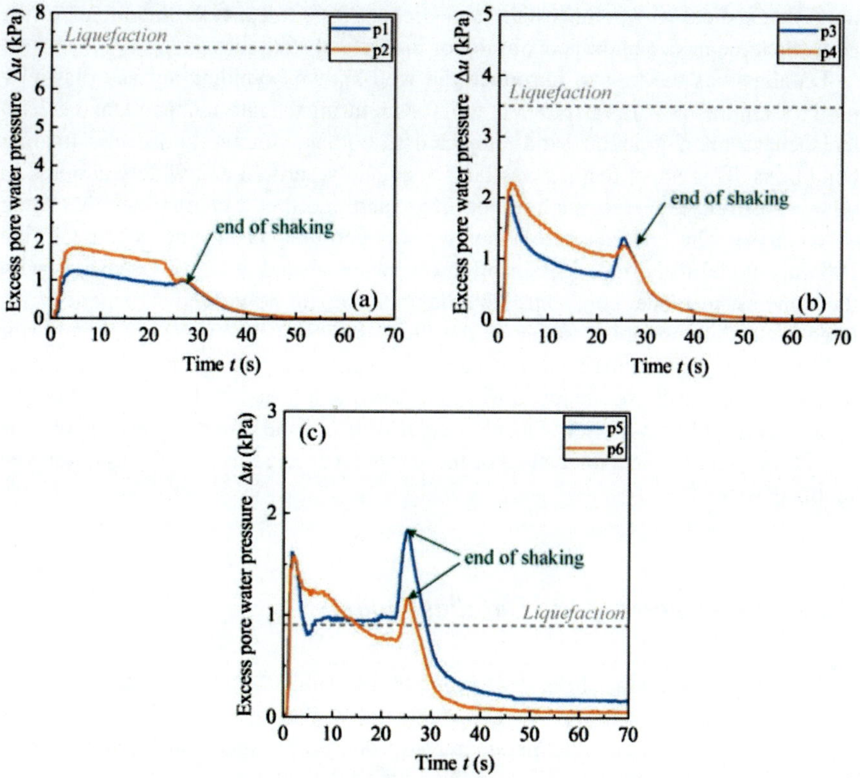

Fig. 5.26 Excess pore water pressures during base shaking at: **a** bottom; **b** half-height; **c** near the top free surface (after Percentile filter with points of window of 250 and percentile of 75)

5.2 Modelling of Sand Liquefaction

In recent decades, considerable efforts have been devoted to the study of sand liquefaction. Numerical modelling is an effective method for investigating such complicated problems and is widely adopted by researchers. Numerical solutions are sometimes preferred owing to their flexibility and low cost of mimicking various test conditions. The development of elasto-plastic soil constitutive models that can logically represent the behaviour of soils under complex cyclic loadings has attracted enthusiastic interest for sandy deposits whose liquefaction response in undrained conditions is considered to have a great impact on the security and serviceability of level ground structures and underground lifelines. Numerous constitutive models based on distinct plastic theories have been proposed; for instance, the PDMY02 model (Elgamal et al. 2002; Yang et al. 2008), the UBCSAND model (Beaty and Byrne 2011), the Dafalias-Manzari model (Dafalias and Manzari 2004) and the PM4Sand model (Boulanger and Ziotopoulou 2015), to name a few. Besides, an innovative stress–strain model has been adapted by Yin et al. (2014) based on the micromechanical viewpoint (Chang and Hicher 2005) to consider the impact of the equivalent mean size and the evolution of the critical state line.

Dafalias was the first to introduce the well-known bounding surface plasticity model (Dafalias 1986) as an efficient tool for capturing the rate-independent mechanical behaviour of granular sand subjected to both monotonic and cyclic loading conditions. This novel feature was later adequately verified and widely adopted in other constitutive models such as the PM4Sand and the Tsinghua models (Wang et al. 2014). The critical state behaviour was carefully taken into account in the Dafalias model to unify the description of the same sand at various densities with the same set of model constants. This flexibility could be viewed as beneficial for large-scale shaking table tests since the final density of the model specimen prior to base shaking cannot be well foreseen and precise controlling during the preparation stage can hardly be achieved. Besides, this model was stably implanted on the open-access and user-friendly OpenSees platform, and the user can easily invoke the model for finite element analysis. For these three reasons, this model was employed in this thesis.

5.2.1 Description of the Dafalias Model

In the following section, a brief description of the Dafalias model is given for triaxial conditions, where $p^{'} = (\sigma^{'}_1 + 2\sigma^{'}_3)/3$, $q = \sigma^{'}_1 - \sigma^{'}_3$ and $\eta = q/p^{'}$. The critical state is computed through the state parameter ψ proposed by Been and Jefferies (1985) to consider the dependency of sand behaviour on the current state:

$$\psi = e - e_c \tag{5.1}$$

where e denotes the current void ratio and e_c is the critical void ratio. The power equation proposed by Li and Wang (1998) is used to depict the relationship between the critical void ratio e_c and mean effective stress $p^{'}$:

$$e_c = e_0 - \lambda_c \times (p_c/p_{at})^{\xi} \tag{5.2}$$

where e_0 is the void ratio with p_c equal to 0, p_{at} is the atmospheric pressure of 100 kPa, λ_c and ξ are two model constants. In terms of elastic strain component (denoted by superscript "*e*") and irreversible plastic strain component (denoted by superscript "*p*"), the incremental stress–strain relationships are determined by the following equations:

$$d\varepsilon_q^e = \frac{dq}{3G}; \quad d\varepsilon_v^e = \frac{dp'}{K} \tag{5.3}$$

$$d\varepsilon_q^p = \frac{d\eta}{3G}; \quad d\varepsilon_v^p = d\left|d\varepsilon_q^p\right| \tag{5.4}$$

where G/K are elastic shear/bulk moduli, H stands for the plastic hardening modulus and d denotes the dilatancy of sand. According to Richart et al. (1970) and Li et al. (1999), G/K can be given by the following equations:

$$G = G_0 p_{\mathrm{at}} \frac{(2.97 - e)^2}{1 + e} \cdot \left(\frac{p}{p_{\mathrm{at}}}\right)^{1/2} \tag{5.5}$$

$$K = \frac{2(1 + \upsilon)}{3(1 - 2\upsilon)} \times G \tag{5.6}$$

where G_0 is a model constant and ν denotes the Poisson's ratio. Generally speaking, sand deformation leading to ultimate rupture is caused by (i) grain sliding and (ii) grain crushing. However, grain crushing only occurs when an extremely high consolidation pressure is employed so as to mimic an extremely deep soil element. Under this assumption, the yield surface (see Fig. 5.27) in the Dafalias model appears as an open wedge which follows the basic assumption that the soil response is governed by grain sliding and grain crushing is only of secondary importance:

$$f = |\eta - \alpha| - m \tag{5.7}$$

where α and m are stress-ratio quantities; therefore, the wedge opens a value equal to $2mp$. In order to meet the stress-ratio dependency, a simple linear relationship can be considered to quantify the plastic modulus H for simplicity:

$$H = h \cdot \left(M^{\mathrm{b}} - \eta\right) \tag{5.8}$$

$$h = \frac{b_0}{|\eta - \eta_{\mathrm{in}}|} \tag{5.9}$$

$$b_0 = G_0 h_0 (1 - c_{\mathrm{h}} e) (\frac{p}{p_{\mathrm{at}}})^{-1/2} \tag{5.10}$$

In this model, the following exponential relations are adopted to relate the bounding and dilatancy stress ratios to the critical state in order to unify the model input constants:

$$M_{\mathrm{b}} = M \cdot \exp(-n^{\mathrm{b}}\psi); \; M_{\mathrm{d}} = M \cdot \exp(-n^{\mathrm{d}}\psi) \tag{5.11}$$

where n^{b} and n^{d} are two positive model constants. Finally, the sand dilatancy is written as follows so as to respect the essence of Rowe's dilatancy law (1962): the dilatancy of granular material should become proportional to the difference of current stress ratio η from that at the phase transformation state:

$$d = A_{\mathrm{d}} \cdot (M_{\mathrm{d}} - \eta) \tag{5.12}$$

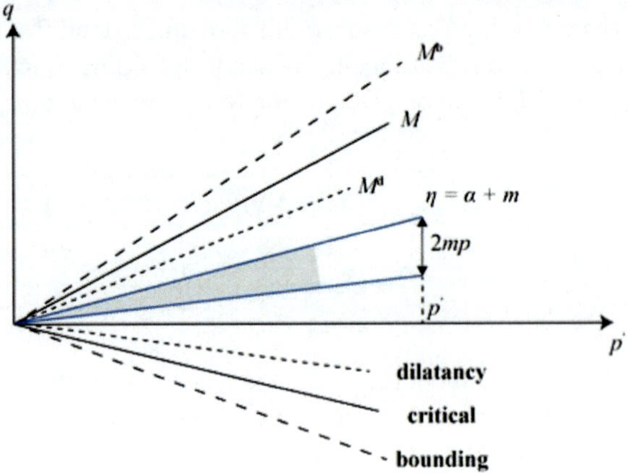

Fig. 5.27 Schematic of yield, critical, dilatancy and bounding lines in triaxial space (after Dafalias and Manzari 2004)

5.2.2 *Performance of the Dafalias Model*

The predictive capabilities of different constitutive models have currently been verified by Ramirez et al. (2018). The objective was to examine whether each constitutive model could yield consistent results (including the generation of excess pore water pressure, the soil settlement during base shaking, etc.) when blindly predicting seismic site performance in a centrifuge test with the model parameters calibrated based on elementary laboratory tests such as undrained/drained monotonic or cyclic triaxial tests.

For this purpose, PDMY02 and SANISAND models were selected. SANISAND is the name for a family of simple anisotropic SAND constitutive models. Each family member was extended from the same original framework developed by Manzari and Dafalias (1997). The subsequent models were modified from the same cornerstone by introducing different ingredients to boost the ability to fit different types of soil behaviour. The version of Dafalias and Manzari (2004) accounting for soil fabric change with fewer model parameters was used in the work conducted by Ramirez et al. (2018) because of simplicity. Figure 5.28 represents the calibration results of different models based on the same experimental result (drained monotonic triaxial test) with the Ottawa sand F65 in a medium-dense state of $D_r = 40\%$. It can be seen that the stress–strain curves at three different consolidation pressures were generally in good agreement with the experimental results. However, none of the three models was capable of mimicking the volumetric strain with the progress of shear. The same Dafalias model in two platforms (OpenSees: https://opensees.berkeley.edu/ and FLAC: https://www.itascacg.com/software/FLAC) provided less contractive responses in the initial phase before attaining the PTS state; subsequently, the later dilatancy response after the PTS state was largely overestimated. Unfortunately, the PDMY02 provided almost tragic results beyond the extent of modern comprehension of basic soil mechanics, even at first glance. Figure 5.29 presents the comparisons between the experimental and numerical results of an undrained cyclic triaxial test (in strain-controlled mode with axial strain amplitude of 0.21%) of the same sand in a dense state of $D_r = 90\%$. In terms of the stress–strain curve in the PDMY02 model, a symmetric pattern is observed, totally different from that sketched in the experimental result. For the Dafalias model, the stress–strain curve developed in a correct non-symmetric manner, especially using OpenSees; namely, the resistance to deformation should be much more prevailing in compression half cycles than extension half cycles. In terms of the generation of excess pore water pressure in the PDMY02 model, an initial dilatancy was observed, whilst a steady contraction response was suggested by the real test. For the Dafalias model, the evolutions of excess pore water pressure in the two platforms (OpenSees and FLAC) were globally consistent with the real test, despite some errors in the first half cycle. In terms of effective stress path in the PDMY02 model, it initially moved to the right, indicating a significant dilatant behaviour. More fatally, the effective stress path could not completely move to the origin point, whilst the experimental result suggested that the test specimen was indeed liquefied and developed a flow-type response, making both the deviator and mean effective stresses infinitely close to zero. Fortunately, the Dafalias model provided a qualitatively correct result even though the initial contraction unfitted the real test (e.g., overestimation in OpenSees but underestimation in FLAC).

Figure 5.30 presents the comparison between the numerical simulation results obtained with the different aforementioned models and the centrifuge experimental test. The model specimen was subjected to a Kobe-L seismic wave at its bottom and the *g*-level was equal to 70 g throughout the test. During base shaking, 4 arrays of transducers for displacement (LVDT), acceleration and pore water pressure measurements were buried in the model specimen covering a wide range from the bottom to the free surface.

The excess pore water pressure histories are shown in Fig. 5.30b. It can be seen from the graph that the generation of excess pore water pressures at shallow depth (near the free surface at $z = 2$ m) and at the bottom ($z = 13$ and 18 m) were largely underestimated. For intermediate layers ($z = 5$, 8 and 13 m), the rapid build-up of excess pore water pressures could be correctly reproduced. However, all three models dissipated the excess pore water pressure much earlier than the real test, especially using the PDMY02 model. The settlement histories at different depths are given in Fig. 5.30c. All numerical simulations provided erroneous results, especially for the PDMY02 model where the volumetric strain was too low to be observed. In the published literature, different investigators have attributed this shortcoming to different possible arguments: (i) Shahir et al. (2012) thought that the use of initial permeability was the prime suspect to explain such an underestimation of settlement during base shaking; (ii) Zhang and Wang (2012) argued that the current U-p formulation for the solid–fluid fully coupled element employed in the finite element analysis was the prime suspect as the velocity effect of pore water was totally neglected.

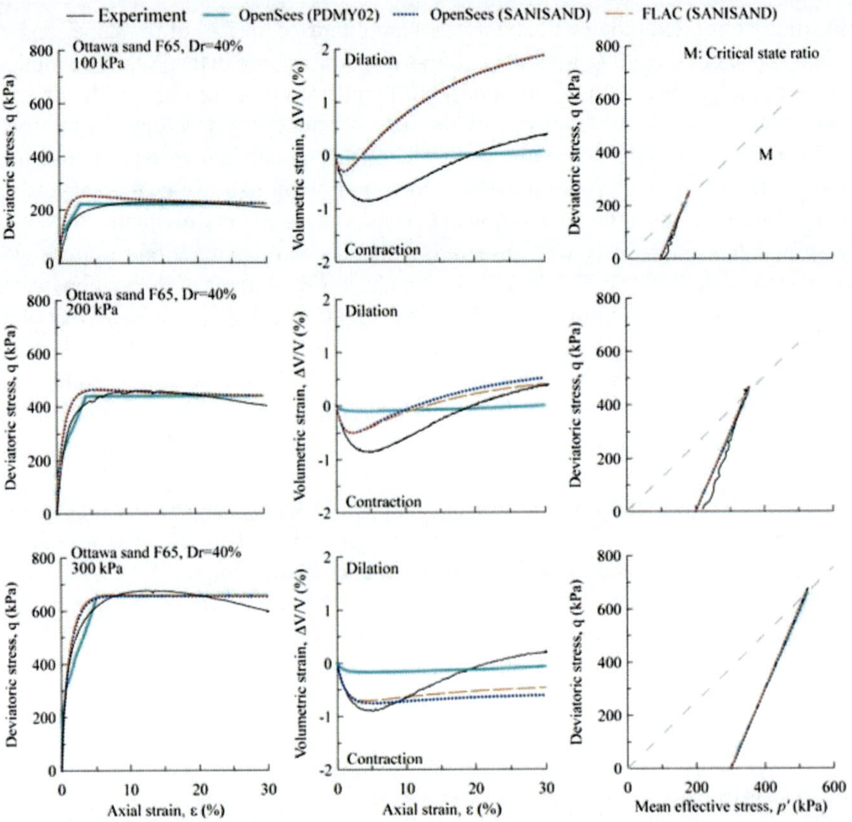

Fig. 5.28 Comparisons between experimental drained triaxial test and numerical simulations for Ottawa sand F65 with $D_r = 40\%$ (from Ramirez et al. 2018)

5.2.3 Calibration Method Using Swarm Intelligence

5.2.3.1 Overview of the Artificial Intelligence Method

As discussed in the above section, the predictive abilities of constitutive models decisively depend on the precise calibration of model parameters. Usually, researchers develop their own methodologies to find the model parameters based on laboratory elementary experiments: involving (i) drained/undrained monotonic/cyclic triaxial tests and (ii) oedometer tests. However, certain fitting parameters in advanced constitutive models might not be, with ease, calibrated with real experimental measurements and often follow a "trial-and-error" procedure where human visual inspection is commonly adopted. With the quick development of information technology in the twenty-first century, artificial intelligence (AI) has become very helpful in shaping the future of material science (Armaghani et al. 2019; Tang and Na 2021). Pertaining

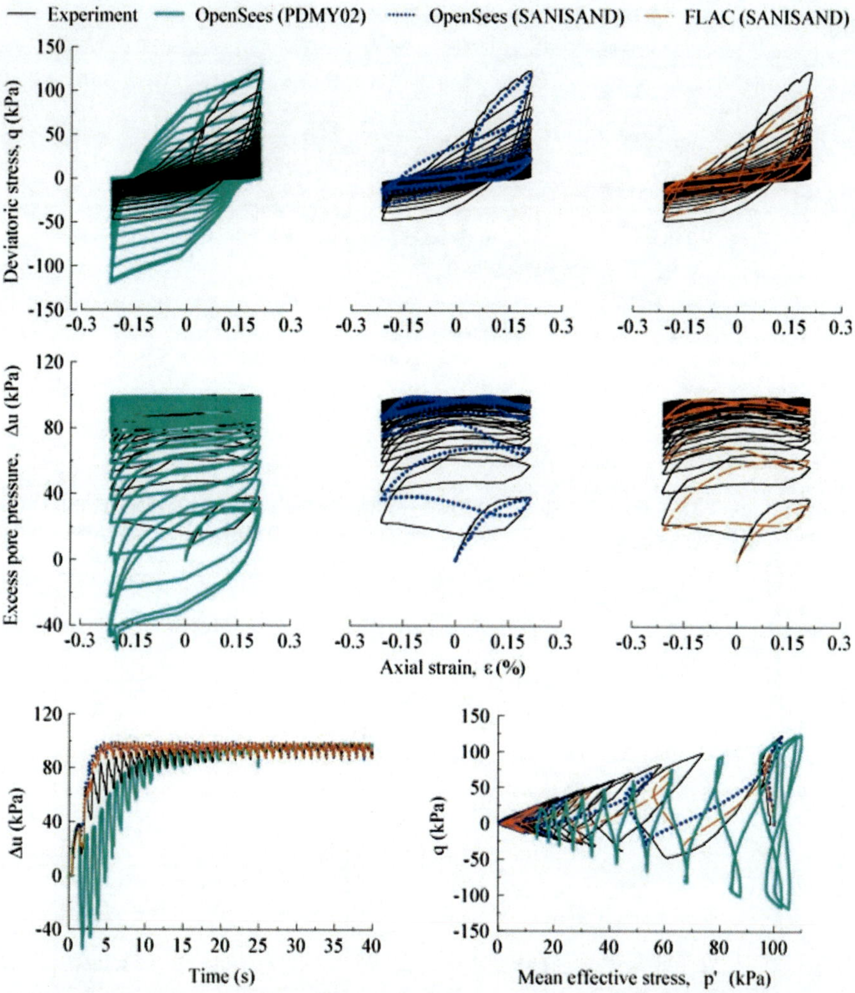

Fig. 5.29 Comparisons between experimental and numerical simulations for Ottawa sand F65 with $D_r = 90\%$ and cyclic shear strain amplitude of 0.315% (or axial strain amplitude of 0.21%) (from Ramirez et al. 2018)

to AI, heuristic algorithms have been brought into the process of model calibration with the purpose of avoiding manual subjective errors. The principle of the method is to interact with a given measure of error so as to search for the pertinent parameters. The key feature as a metaheuristic is that it can search a very large space of candidate solutions without making additional assumptions about the problem itself being optimized. As a result, the partial gradient with respect to each fitting parameter is not required when the combined influence of fitting parameters needs to be fully considered. This is quite different from the classic optimization methods such as the gradient descent and quasi-newton methods. In the pioneering work of Pal

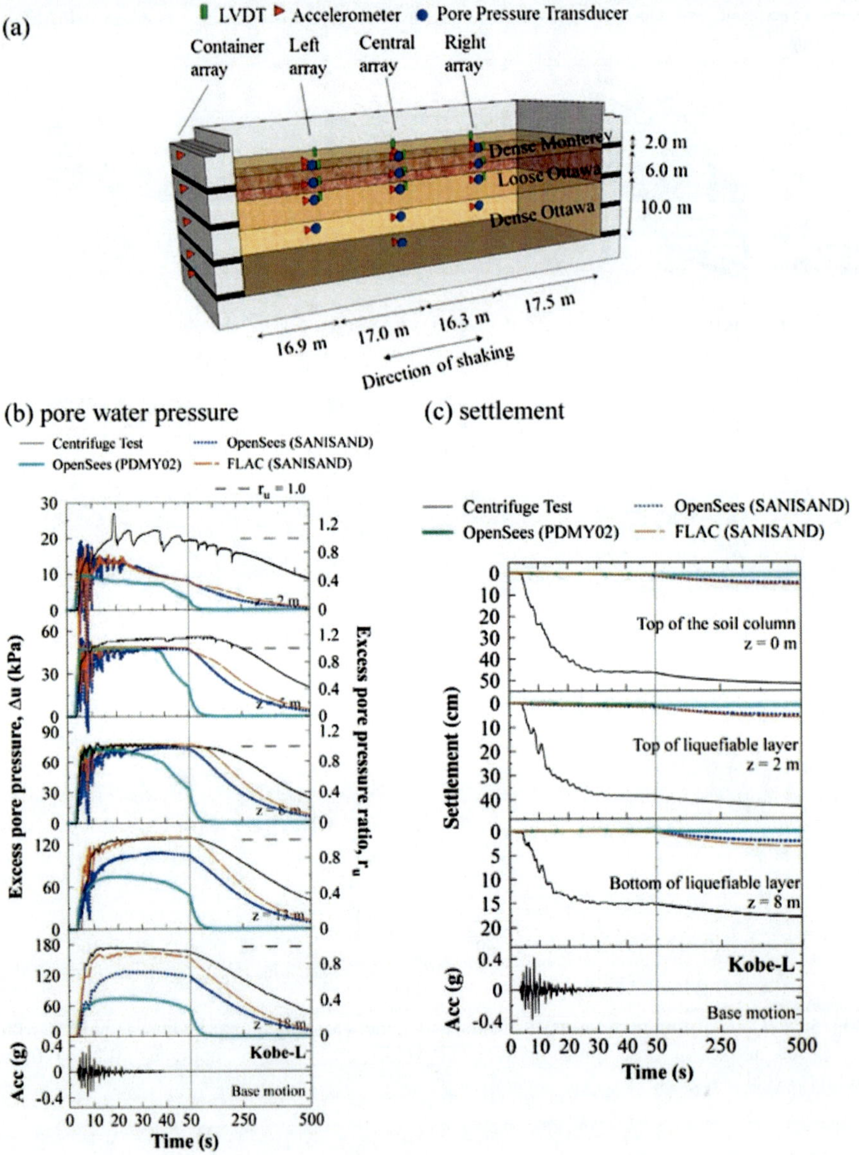

Fig. 5.30 Comparison between experimental and numerical simulations for a centrifuge test: **a** instrumentation layout; **b** pore water pressure histories at different depths; **c** settlement histories at different depths (after Ramirez et al. 2018)

et al. (1996), the genetic algorithm (GA) was selected to fit the Hierarchical Single Surface model. The main advantage (Yeh et al. 2020; Mendez et al. 2021) over the traditional method of using the GA is the ability to take care of the overall quality of the simulated results instead of sequentially finding one or more parameters in the first place to later search the other parameters.

From an engineering point of view, the cyclic triaxial test plays a more prevailing role since it is capable of reproducing a sequence of cyclic deviator stresses on the test specimen, very similar to the alternated wave loading for sand elements in level ground. However, the above successful applications methods were only limited to the monotonic tests, probably owing to the following two independent factors: (i) it is difficult to evaluate the relative error between measured and simulated results since a strain value in a cyclic triaxial test can map various stress values; (ii) with the incorrect model parameters, especially during the initial few iterations, the simulation of a cyclic triaxial test is much more prone to divergence, making the calculation extremely time-consuming and even sometimes impossible. In order to overcome these problems, this section presents a novel calibration methodology based on the Particle Swarm Optimization (PSO) algorithm. Compared with the GA method, the PSO method is relatively simple since no extra code mimicking the evolution of natural selection is needed, and the method could directly hunt for the optimized values in the proposed search space. A special multi-objective measure of error interacting with the PSO method is presented in this section. It reflects the combined impact of monotonic and cyclic triaxial simulations by giving each the same importance and the clear benefit of such a definition is to guarantee the global quality of both the computed results. Besides, two additional interruption criteria were fully considered for the cyclic triaxial simulation so that the PSO method is prevented from wasting time in searching the unreasonable space of candidate solutions. Finally, an extension example is demonstrated to explore the versatility of the proposed programme and an in-depth analysis of optimization ability is done with the elastoplastic theory.

The target granular material in this study is the HN31 sand, which is characterized by uniform sub-angular grains with a mean grain size D_{50} equal 0.35 mm and a uniformity coefficient C_u equal to 1.57. The Dafalias model involves 15 model parameters in total. For the HN31 sand, the shear modulus constant G_0 is assumed to be equal to 176 (Zhu and Cheng 2020), and the Poisson's ratio $\nu = 0.05$ is widely adopted by many researchers for different sands (e.g., Toyoura, Monterey and Ottawa sands in Dafalias and Manzari 2004 and Ramirez et al. 2018, respectively). The model parameters concerning the critical state can be calibrated with the monotonic triaxial results presented by Benahmed (2001) by performing the optimum fitting procedure (as shown in Fig. 5.31): $\lambda_c = 0.065$, $e_0 = 1.00$ and $\xi = 0.40$. In terms of the effective stress path in the *q-p*' space, the critical stress ratio M_c was equal to 1.37 (Benahmed 2001). Papadimitriou et al. (2001) suggested that the value for the yield surface constant m of the Dafalias model could be considered to be of the order of $M_c/100$, namely $m = 0.014$ in the present work. The critical state stress ratio c between compression and extension has been experimentally found to be equal to 0.712 (Benahmed 2001). The bounding/dilatancy parameters n^b/n^d were computed following the empirical equations proposed by Li and Dafalias (2000), which were estimated to be 0.60 and 2.50, respectively. It should be mentioned that different experimental data might probably give different values for n^b and n^d; thus, a large number of real tests were averaged to obtain the real values.

All the above model parameters (the 10 parameters listed in Table 5.3) can be, with ease, determined using well-documented experimental data. However, the other 5 parameters are suggested to follow a "trial-and-error" procedure (Dafalias and Manzari 2004), whilst these parameters certainly play a vital role in controlling the computed liquefaction response, especially the two plastic moduli (h_0 and c_h) and the dilatancy parameter A_0 directly linking to the contraction behaviour and the generation of the excess pore water pressure. As a consequence, the fitting of these 5 parameters was done by the PSO procedure with no need for manual adjustments, which will be detailed in the next section.

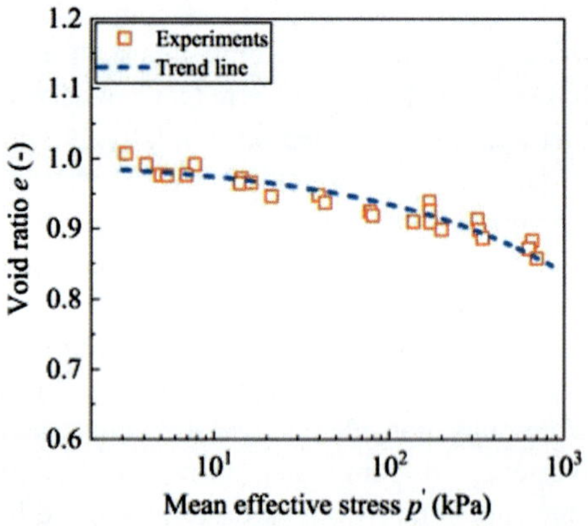

Fig. 5.31 Calibration of the critical state line (from Benahmed 2001)

Table 5.3 Model parameters of the Dafalias model for HN31 sand

Constant	Parameter	Value
Elasticity	G_0	176
	ν	0.05
Critical state	M_c	1.37
	c	0.712
	λ_c	0.065
	e_0	1
	ξ	0.4
Yield surface	m	0.014
Plastic modulus	n^b	0.6
Dilatancy	n^d	2.5

5.2.3.2 Methodology Based on the PSO Method

Originated from Kennedy and Eberhart (1995), particle swarm optimization (PSO) in computational science intended to stylize the social behaviour of birds in the foraging process by having a population of candidate solutions, here dubbed particles. Each particle is treated as a single point in the user-defined D-dimensional space. With regard to a given measure of error, each individual could adjust its flying tendency as a function of its own flying experience and its companions' flying experience. During the iteration $k + 1$, the velocity v_{id}^{k+1} and position x_{id}^{k+1} vectors of the particle i are updated according to the following mathematical formulas:

$$v_{id}^{k+1} = \omega \cdot v_{id}^{k} + c_{\mathrm{p}} r_1 \cdot (Pbest_{id}^{k} - x_{id}^{k}) + c_{\mathrm{g}} r_2 \cdot (Gbest_{d}^{k} - x_{id}^{k}) \tag{5.13}$$

$$x_{id}^{k+1} = x_{id}^{k} + v_{id}^{k+1} \tag{5.14}$$

where ω denotes a positive inertia weight; thus, the first part in Eq. (5.13) represents the ability of each particle to maintain its previous velocity (namely, the "inertia effect" in the physical world). The c_{p} in the second part is termed as a cognitive parameter and r_1, r_2 are two random numbers $\in [0, 1]$ to involve randomness in the calculation. The *Pbest* represents the optimal position searched by an individual particle till the iteration number equals to k so that the second part in the Eq. (5.13) can automatically consider the impact of local optimal solution while updating the velocity vector. Similarly, the c_{g} in the third part is termed as a social parameter and *Gbest* denotes the optimal position searched by the particle swarm so that the impact of global optimal solution is fully considered while updating the velocity vector.

For a given problem, the choice of PSO parameters could have a significant effect on the subsequent performance; thus, the selection of these parameters has become the primary objective of much research (among Bratton and Blackwell 2008; Eberhart and Shi 2000; Shi and Eberhart 1998; Trelea 2003). Generally, an inertia weight $\omega = 0.80$ is convictive when lacking the knowledge regarding the treated problem, as suggested by Shi and Eberhart (1998). In order to balance the local and global solutions, c_{p} and c_{g} were set to be the same in the present work, both equal to 0.50. The key issue for the PSO method is the proper definition of the objective function accounting for the error between the numerical and experimental results, and the straightforward goal of the PSO method is to minimize the objective function value to bring the computed results as close to the real experiment as possible. In this study, two standard experimental results were employed to calibrate the 5 target parameters: (i) a drained monotonic triaxial test under isotropic consolidation stress $\sigma'_{\mathrm{c}} = 200$ kPa with the purpose of well capturing the internal friction angles of granular material and (ii) an undrained cyclic triaxial test (cyclic stress ratio $CSR = \tau/\sigma'_{\mathrm{c}} = 0.22$, loading frequency $f = 0.10$ Hz and $\sigma'_{\mathrm{c}} = 200$ kPa) to consider the cyclic behaviour of granular material. It should be mentioned that the employed cyclic triaxial result was seriously double-checked by a repeatability test (Zhu et al. 2021b). As already known to all, the evolution of volumetric strain during shearing in the case of drainage is also an important characteristic associated with the inherent dilatancy of granular materials. Nevertheless, when using the finite element method, the prediction is most likely an underestimation (Howell et al. 2015), as discussed in section §5.2.2. The effectiveness of the finite element method itself towards liquefaction analysis is out of the scope of this section; therefore, the volumetric deformation during shearing is not considered in the identification process.

Based on the above elements, the multi-objective function in this section is defined as the sum of errors with respect to both aforementioned experiments:

$$Obj = \text{Part1} + \text{Part2} = W_{\text{momo}} \cdot RSS_{\text{mono}} + W_{\text{cyc}} \cdot RE_{\text{cyc}} \quad (5.15)$$

where Part1 considers the normalized residual sum of square (*RSS*) of stress–strain curve with respect to the drained monotonic triaxial test, and Part2 considers the relative error with respect to *the number of cycles* N_{cyc} *attaining the initiation of sand liquefaction.* Under the undrained triaxial condition, the excess pore water pressure ratio ($r_u = \Delta u/\sigma'_c$) is considered to be a fairly suitable indicator to estimate the cyclic behaviour of sand liquefaction. A typical triaxial test associated with HN31 sand liquefaction (Zhu et al. 2021b) is displayed in Fig. 5.32. Two phenomena could be observed when r_u became equal to 0.80 during the number of cycles N_{cyc} between 13.5 and 14.5: (i) the transient axial deformation began to occur in Fig. 5.32a and (ii) a "reversed-z" stress–strain curve became increasingly more obvious in Fig. 5.32b. Both the phenomena were referred to the outset of sand liquefaction (Xie 2011); hence, the r_u value of 0.80 was taken as a triggering liquefaction threshold. Besides, this parameter is naturally dimensionless. It is of cardinal importance to mention that the conventional *RSS* in Part1 was intentionally normalized by the consolidation pressure σ'_c so that it would be unit-independent for the suitable scaling with the addition of Part2 in the equation. In general, these two parts are naturally of different orders of magnitude. Two weight factors $W_{mono} = 10$ and $W_{cyc} = 1$ were then applied to balance contributions and gain a better global quality for both computed results.

In order to efficiently reproduce the computed results of the Dafalias model, two solvers were respectively implemented in the Python environment in order to simulate the mechanical response of a unit element (thanks to the OpenSees platform), using the Numpy (http://numpy.org) and the OpenSeesPy library (https://openseespydoc.readthedocs.io/en/latest/). After choosing the fitting parameters, the PSO programme was initiated by randomly placing a population of particles in the provided search space. During each iteration loop, two Python solvers produced the computed results of the Dafalias model to obtain the final value of the objective function one particle after another (see Fig. 5.33a). As mentioned before, the simulation of cyclic triaxial tests is more prone to divergence issues, especially during the initial few iterations, making the calculation time extremely long and even sometimes impossible. Thus, two interruption algorithms were fully designed for each calculation step. First, once the nodal response of accumulated excess pore water pressure reaches the r_u value of 0.80, the current simulation is automatically stopped in order to escape the time-consuming post-liquefaction phase since the obtained numerical result was already sufficient to deduce the value of Part2. Second, for each calculation step, the time step

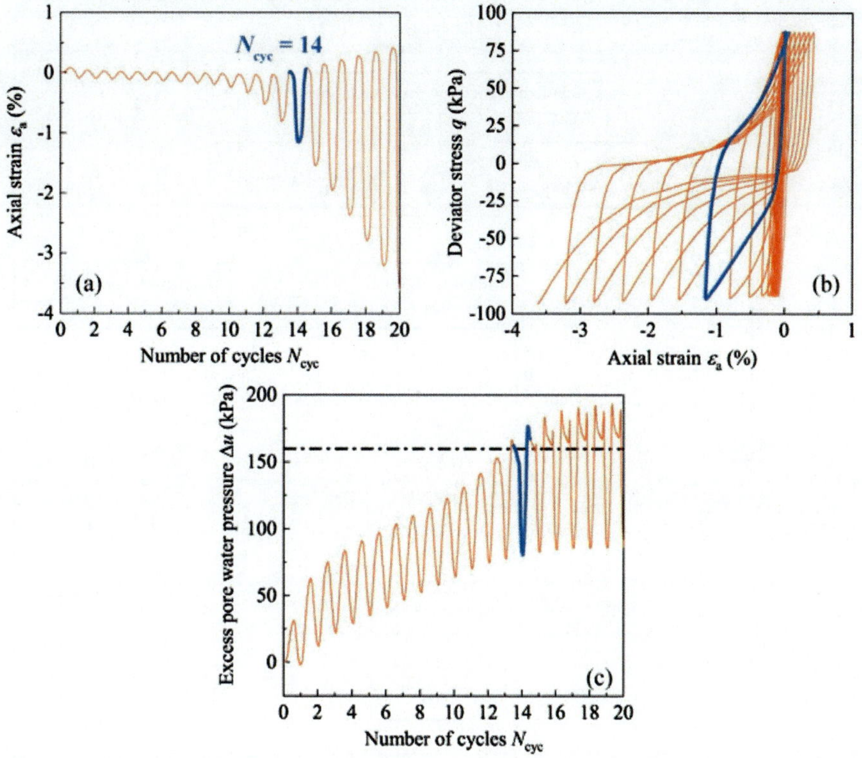

Fig. 5.32 Typical triaxial test involving sand liquefaction in terms of **a** axial strain versus number of cycles; **b** stress–strain curve; **c** excess pore water pressure versus number of cycles

Δt was set in an adaptive manner, reducing it possibly by half in case of divergence. A huge value of 9999 (see Fig. 5.33a) will be directly assigned to the Part2 of the objective function if the current step still diverges after three reductions in a row. During the iteration, as long as the particles search the unreasonable space of candidate solution leading to the convergence problem (see Fig. 5.33b), they certainly have no chance of becoming the optimal position of the swarm (*Gbest* in Eq. 5.13) with the intentional punishment value of 9999. The goal of this mechanism is to speed up the calibration programme by punishing the particles within the unreasonable search space and to force them to leave the wrong area quickly. For the same reason, if there is no liquefaction initiation characterized by the computed r_u (always remaining lower than 0.80) throughout the numerical simulation, the value of Part2 will directly be supposed to be 99 (see Fig. 5.33c). Finally, the velocity and position vectors of particles could be updated for the next iteration, according to Eq. (5.13) and Eq. (5.14).

Although the calibration procedure can be totally entrusted to the PSO procedure, the search space is still man-made and thus subjective. The proper definition of such a space is very important and, sometimes, requires some experience. Generally

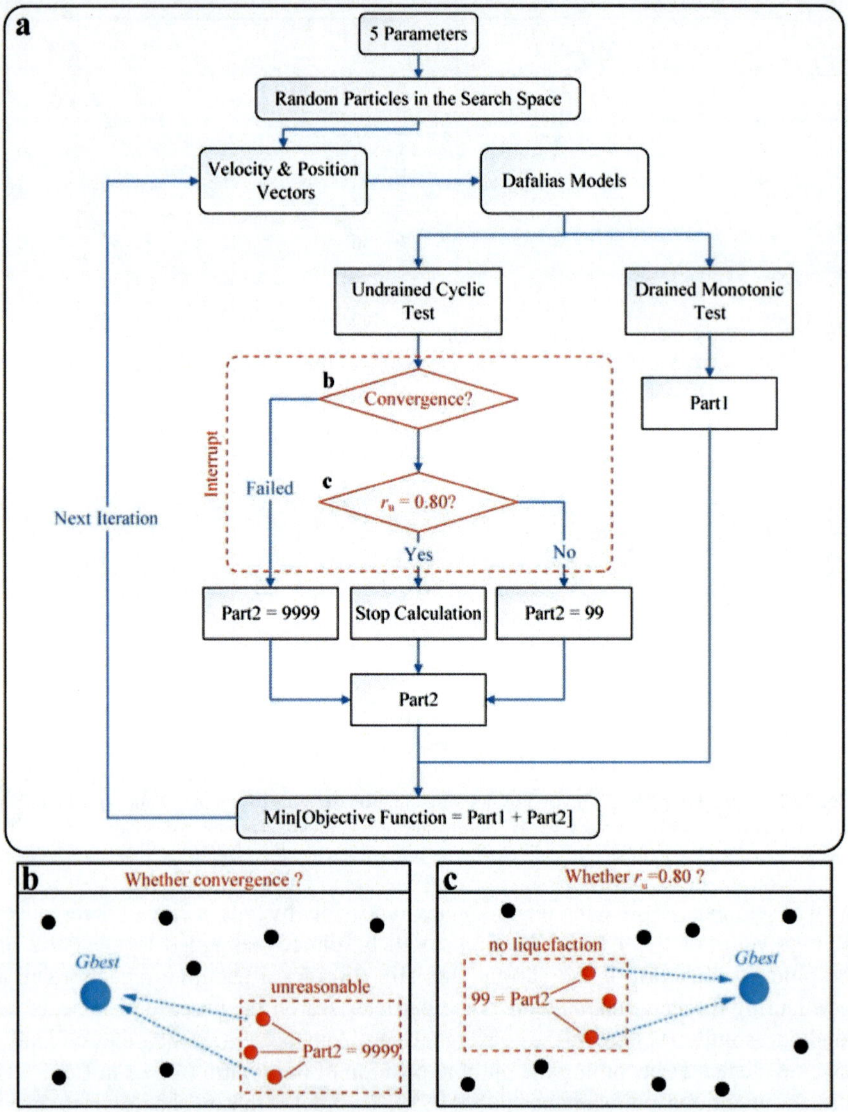

Fig. 5.33 **a** Flow chart with the Python solver to compute the value of the objective function; **b** interrupt mechanism for convergence; **c** interrupt mechanism for liquefaction judgement

speaking, enlarging the space might drastically increase the number of iterations and decrease the convergence rate in most cases. In contrast, decreasing such a space certainly reduces the opportunity to find the optimum set of parameters. Thus, the volume of the search space needs to be seriously examined after the calibration procedure: (i) when it is too small, all particles are highly tended to cluster on the boundary of the employed search space; (ii) when it is too big, a huge difference in

Table 5.4 Search space and the calibrated values for HN31 sand

Constant	Parameter	Range	Calibrated value
Plastic modulus	h_0	[2,30]	7.224
	c_h	[0.30,1.30]	1.052
Dilatancy	A_0	[0.05,0.80]	0.087
Fabric-dilatancy parameters	z_{max}	[10,100]	76.363
	c_z	[400,800]	511.838

computed results might be found; thus, different particles might probably cluster on various points in the provided search space.

5.2.3.3 Effectiveness of the PSO Method

(i) **Assessment of optimization ability using HN31 sand**

The employed search space and the calibrated values are tabulated in Table 5.4. It can be seen that all 5 target parameters fell well within the proposed range, indicating that such a space was correctly imagined for the HN31 sand. Figure 5.34 depicts the initial positions of all particles with a population of 10 and maximum iteration number of 200.

Before the first iteration in Fig. 5.34a, 10 particles were randomly placed in the proposed search space. And near the final stage, at the iteration 195, they were highly tended to cluster at a single point. The coordinates of this point could be believed to be the optimized model parameters. In addition, the best value of objective function till the current step was normalized by its peak value in Fig. 5.34 to display its intrinsic evolution distinctively. During the progression of iterations, the initial high value steadily decreased to be much smaller. After the iteration 50, no clear benefit could be further gained, indicating that the employed maximum iteration number of 200 was sufficient for the calibration purpose. By combining all the above phenomena, the obtained parameter values after the PSO procedure should be deemed to be meaningful to reproduce the mechanical behaviours upon shearing of the HN31 sand using the Dafalias model.

Figure 5.35 presents the comparisons between the computed and experimental results for the HN31 sand in a monotonic loading condition. The stress–strain curves described in Fig. 5.35a demonstrate that not only the reference test (200 kPa) but also other experimental results (100 kPa and 400 kPa) could be successfully reproduced with the calibrated model parameters using the PSO method. Besides, the effective stress paths drawn in Fig. 5.35b show that the correct internal friction angles could be derived at different shearing stages (e.g., maximum ϕ'_{max} and critical ϕ'_{cri} states), as compared with the real experiments.

Figure 5.36 presents the comparison between the computed and experimental results for the HN31 sand in a cyclic loading condition. The excess pore water pressure Δu against the number of cycles N_{cyc} curves drawn in Fig. 5.36a prove that

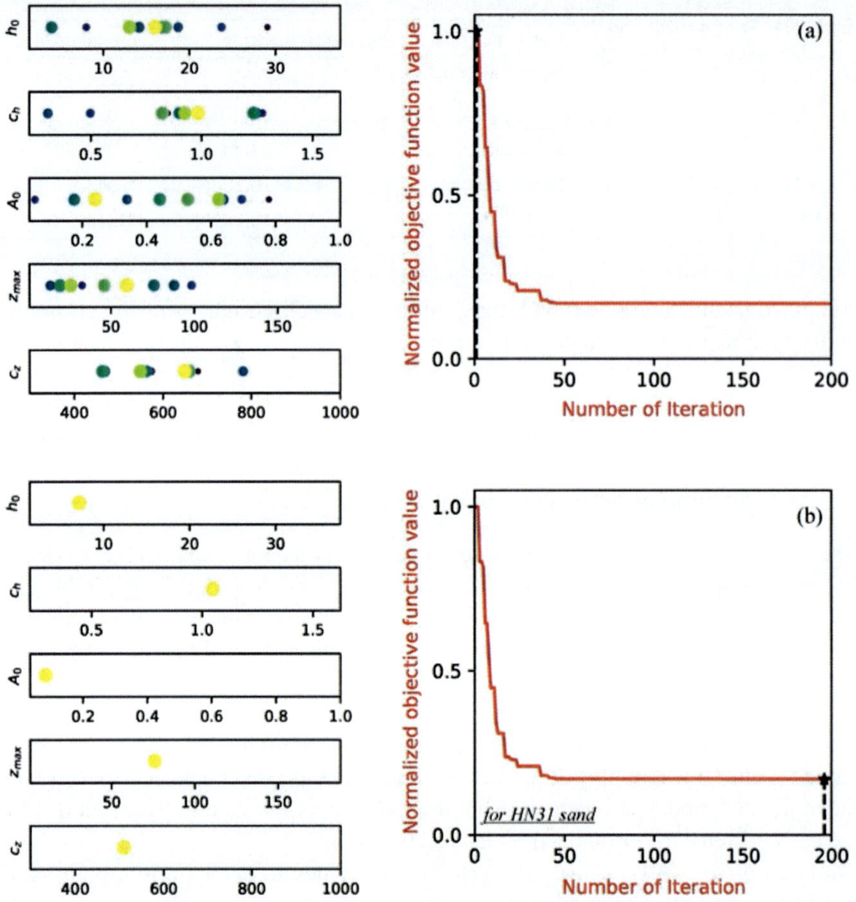

Fig. 5.34 Particle iterative positions and the normalized objective function **a** at the first iteration; **b** near the last iteration

the computed result conformed well to the given criterion of $r_u = 0.80$ since only a minor variation in N_{cyc} could be observed for the achievement of this criterion. After the outset of sand liquefaction, an apparent decrease in Δu could be found for the computed results, different from the experimental reference. This shortcoming is logical since the sand material passed from the solid to the liquid phase at this experimental stage; and this effect is not numerically considered in the Dafalias model. The effective stress paths drawn in Fig. 5.36b show that the two curves were almost identical. Provided that the objective function (Part2) accounting for the relative error for such a test was based on N_{cyc}, the PSO method indeed made a compromise by destroying to some extent the initial contraction tendency (as outlined in Fig. 5.36b), allowing a pertinent value of this indicator to be successfully reproduced for the computed result.

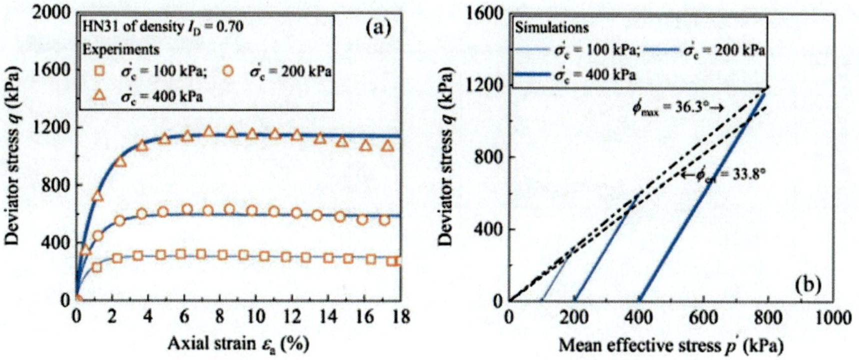

Fig. 5.35 Comparisons between the computed (in blue) and experimental results (scatter points from Benahmed 2001) in monotonic triaxial test: **a** stress–strain curves; **b** effective stress paths

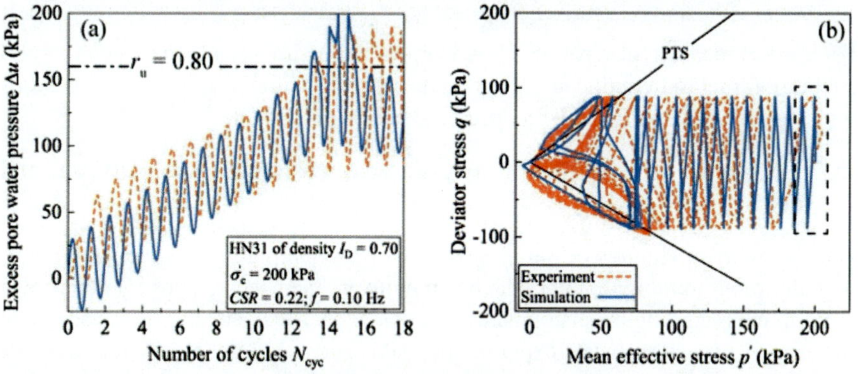

Fig. 5.36 Comparison between the computed (in blue) and experimental results in cyclic triaxial test: **a** excess pore water pressure versus number of cycles curves; **b** effective stress paths

In order to further explore the performance of the proposed calibration procedure, another undrained cyclic test with a different loading mode (displacement-controlled mode with peak axial strain value ε_a equal to $\pm 0.40\%$) was used to blindly examine the calibrated model parameters. Figure 5.37a presents the comparison with respect to the stress–strain curves between the computed and experimental results. The computed curve is globally in good agreement with the experiment despite some small errors could be found in the compression half cycles. In particular, the computed stress–strain curve developed in a correct asymmetric manner, conforming to the mechanical characteristics of granular materials. Namely, the resistance to compressive shearing should be much more prevailing than that to extensional shearing. This mechanical characteristic could be well reproduced since the evolution of fabric for granular materials is fully considered in the Dafalias model. In terms of the evolution of Δu during shearing, the predictive N_{cyc} attaining the liquefaction triggering r_u of 0.80 was almost the same as the experimental observation, as shown in Fig. 5.37b.

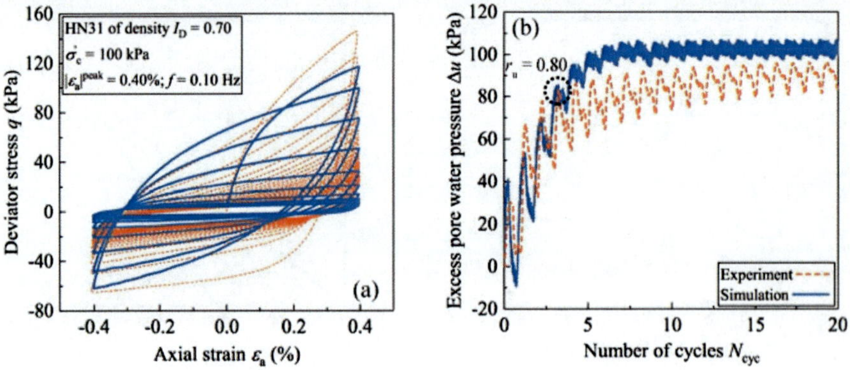

Fig. 5.37 Comparison between the computed (in blue) and experimental results in cyclic triaxial test (displacement-controlled mode): **a** stress–strain curves; **b** excess pore pressure versus number of cycles curves

For a given measure of error, this issue appears to be important since the proposed calibration method continued to work satisfactorily without violating its previous premise despite of being under a different loading condition.

(ii) **Assessment of versatility of the proposed programme using Ottawa sand F65**

One of the main features of the proposed optimization platform is the adoption of modular programming thought which can motivate the users to construct their own programme regarding the encountered problems, only by changing (i) the objective function and (ii) the relevant experimental data. The use of the platform is versatile and should not only be limited to the case discussed above: *optimization based on a monotonic test and a cyclic test*. To further explore this possibility, an extension example is given here and the effectiveness of the PSO method will be assessed with the elastoplastic theory.

As mentioned previously, the Dafalias SANISAND model was intensively applied in a study of site response in a layered liquefiable deposit (Ramirez et al. 2018). With due care and diligence, a series of model parameters was calibrated. Based on these parameters, the idea now appears to be attractive, if the proposed PSO programme can be launched to explore whether these parameters could still be improved. In order to make the difference, the optimization, this time, relied on *three monotonic tests under different consolidation stresses* (Ottwa sand F65 of $I_D = 0.40$ or $e = 0.70$ under 100 kPa, 200 kPa and 300 kPa, respectively). The first part of the objective function was then defined as the sum of three normalized *RSS* of stress–strain curve whilst the second part was removed. The same 5 target parameters were entered in the PSO iteration and the other 10 experiment-determinable parameters were kept constant. It is worth noticing that the normalization of *RSS* remained necessary to ensure the balance of the objective function since the shear strength of sand sharply increases with the consolidation stress.

The initial positions of all particles with a population of 10 and the applied search space are displayed in Fig. 5.38. It can be seen that the particles finally stayed at a single point away from the boundaries that specifies a proper consideration of such a space for this sand.

The experimental and computed stress–strain curves are presented in Fig. 5.39 with a local enlarged version below where two different phenomena can be recorded. First, both the computed curves well agree with the experimental results while approaching and reaching the critical state ($\varepsilon_a > 17\%$). Second, as compared with the curves presented by Ramirez et al. (2018) in the initial phase of loading as zoomed below ($0\% < \varepsilon_a < 8\%$), the fitting parameters further optimized by the PSO method, in general, yielded more crooked curves whose responses matched the nonlinear characteristic of sandy specimens better. The evolution of normalized objective function

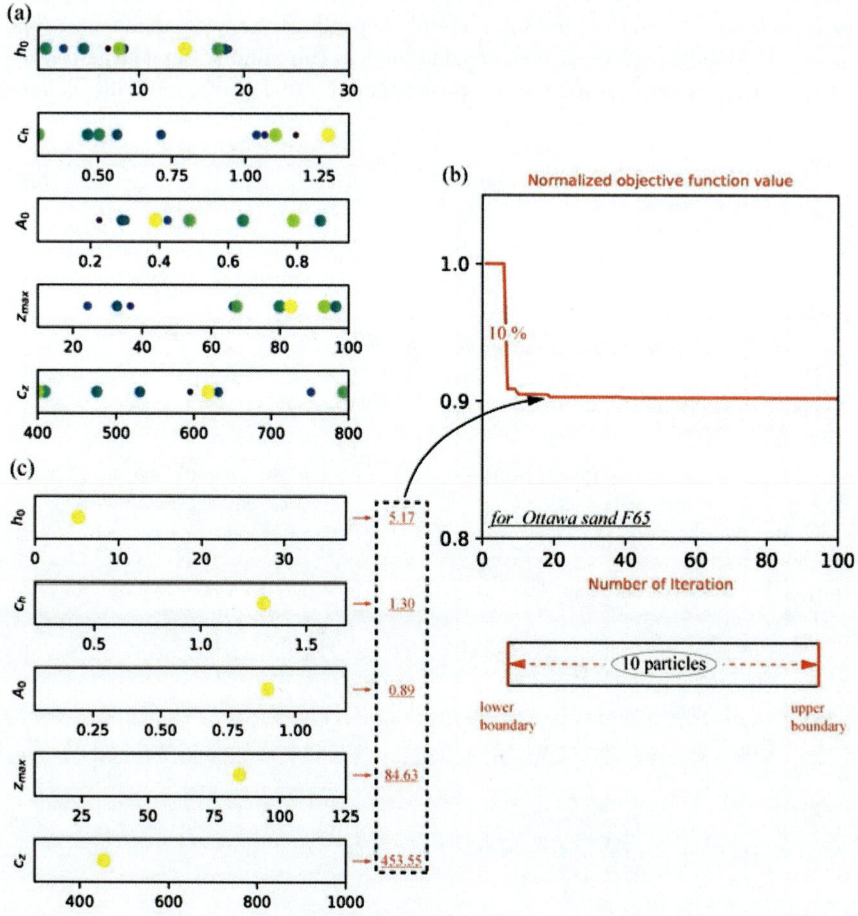

Fig. 5.38 Particle iterative positions at **a–c** first and last iteration; **b** the normalized objective function

is displayed in Fig. 5.38b and it is gratifying to admit that the model parameters initially proposed by Ramirez et al. (2018) were already of quality. However, a limited improvement (approximately 10%) could still be gained.

In statistics, the coefficient of determination, denoted r^2, is a common measure to quantify how well observed outcomes are replicated by the model. Therefore, this indicator was taken to inspect the parameters further calibrated with the PSO method. For the lower consolidation pressures of 100 and 200 kPa, the r^2 presented in Fig. 5.40 reconfirms that the PSO method made a supplementary contribution to the computed results, especially for 100 kPa. However, for the consolidation stress equal to 300 kPa, a minor loss of r^2 (from 0.98 to 0.97) is seen in the graph since a compromised choice was necessary for the PSO method in order that a better global quality with respect to the given objective function could be achieved.

The above compromise made by the PSO method can be explained by the underlying rationale of the elastoplastic theory. Although the algebraic value of plastic modulus H (from Eqs. (5.8) to (5.10)) mainly depends on the two parameters h_0 and c_h, this modulus has a physical meaning, although its calibration cannot be performed from real experiments. Similar to the elastic modulus, the plastic modulus, defined

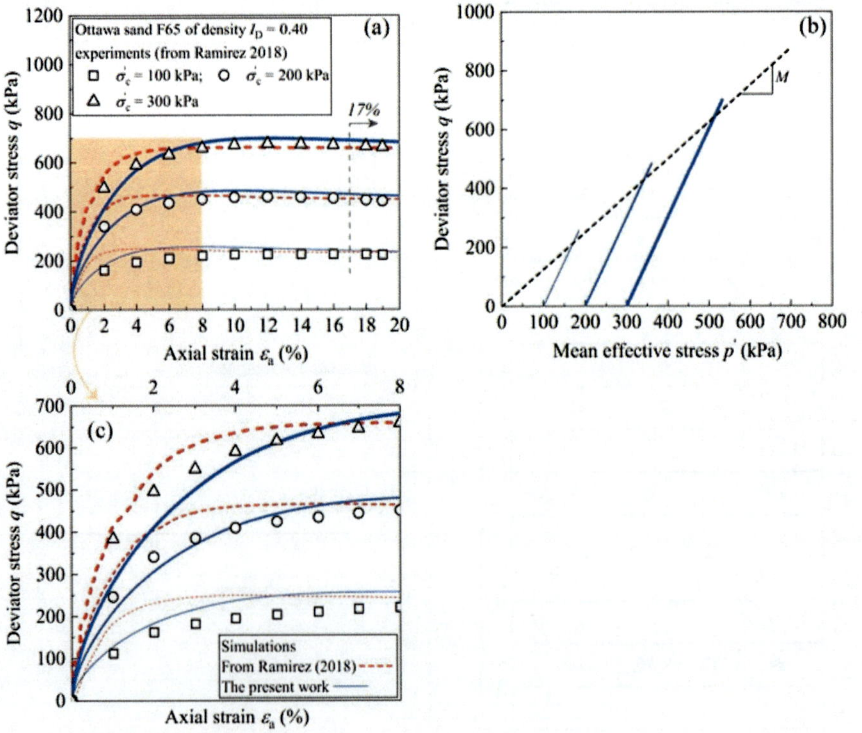

Fig. 5.39 Comparisons between the computed and experimental results in monotonic triaxial test for Ottawa sand F65: **a** stress–strain curves; **b** effective stress paths; **c** local enlarged version

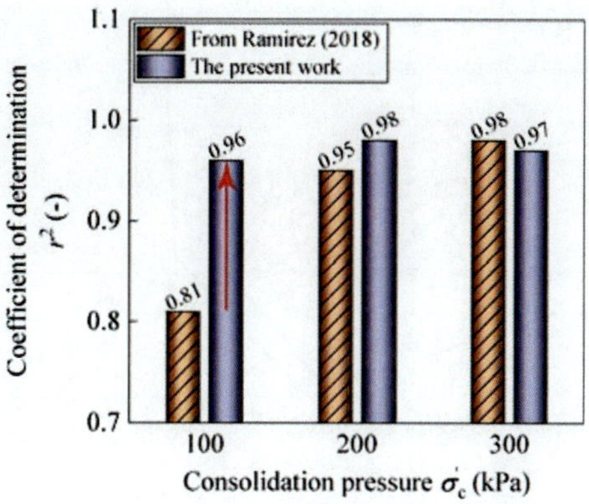

Fig. 5.40 Comparisons of two series of model parameters under different consolidation stresses in terms of coefficient of determination r^2

as the *normal* component of stress increment divided by the *plastic* strain increment (see Fig. 5.41a), serves as an index to describe the extent to which the plastic deformation can be generated while yielding occurs. In other words, there is no plastic deformation when H approaches infinity and only elastic behaviour is exhibited by the model. An example of profiting the above mechanism in the Dafalias model is its ability to mimic the pure elastic response of sandy specimen (as outlined in red in Fig. 5.41b) when loading reversal takes place. While passing through the reversal, the value of η_{in} is set being the same as the current stress ratio η, leading H to be infinite (see Eq. 5.9) and this eradicates the concern to think any plastic deformation.

Figure 5.42 displays the 3D wireframe of the plastic modulus H from different isometric views using two series of model parameters, respectively. In the graph, the modulus of the present work is always lower than the reference, especially for the target density range ($e = 0.70$ in Fig. 5.42b). Now, it should be convenient to understand the choice that the PSO method made as a black box. At the very start of the loading stage, the PSO method attempted to turn down the plastic modulus H.

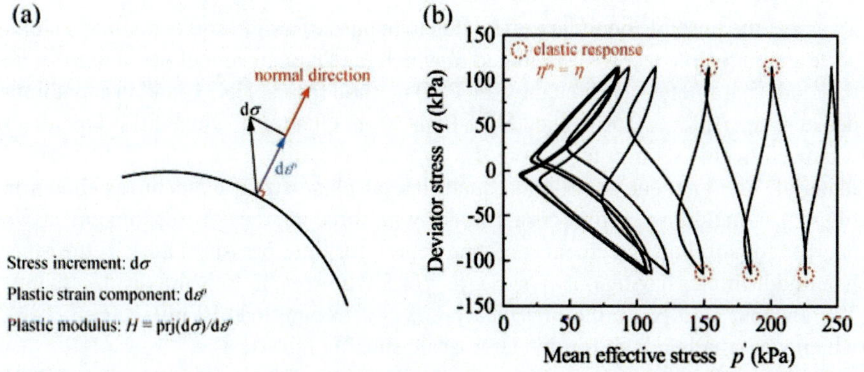

Fig. 5.41 Schematic of **a** the plastic modulus definition; **b** pure elastic responses (after Dafalias and Manzari 2004)

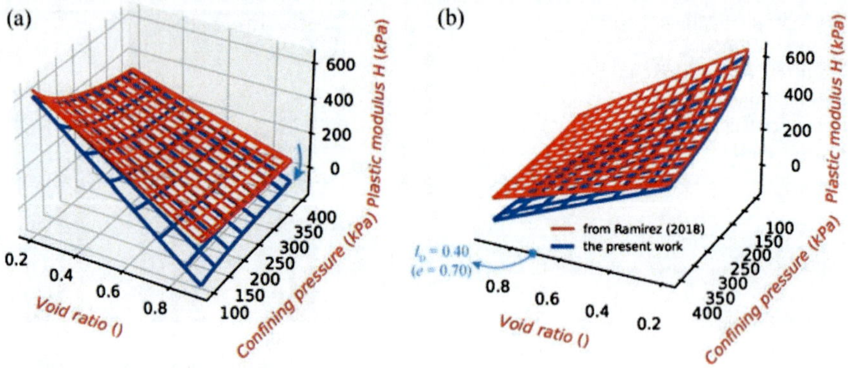

Fig. 5.42 3D wireframes of plastic modulus H from different isometric views

As a result, the development of plastic strain got boosted. This effect inhibited the rapid growth of elastic strain part, which was the key ingredient to yield the stress build-up. Then, more curved results were numerically generated, as clearly shown in Fig. 5.39. However, the above reasoning cannot logically explain what has been observed for the critical state in Fig. 5.39. As the calculation reached the critical state (namely $e = e_c$), the state parameter ψ was set to be 0. From Eqs. (5.8) and (5.11), it can be deduced that the plastic modulus H was automatically zeroed out, irrespective of the algebraic value of h_0 and c_h. Thus, as for the large strain range that the critical state prevailed, the modification of the target parameters through the PSO method was unbeneficial because the parameters governing the critical state had been clearly determined by another method: the optimum-fitting method (see Fig. 5.31).

5.2.4 Numerical Assessment of Boundary Effects

To assess the possible boundary effect due to the use of a rigid soil container, a large-scale shaking table test was simulated in which the Dafalias model was applied with the model parameters calibrated by the PSO method for the HN31 sand in a medium-dense state ($I_{\mathrm{Dmat}} = 0.50$). The "Nine Four Node Quad u-p" element of OpenSees was used in this section to capture the dynamic response of a solid–fluid coupled material. The element is a 9-node quadrilateral plane-strain element (as shown in Fig. 5.43), including (i) four corner nodes with three degrees of freedom each (two degrees for solid displacements and one degree for fluid pressure) and (ii) the other five middle nodes having only two degrees of freedom each, for solid displacements. The element size Δz in the present work was set equal to 0.10 m (see Fig. 5.43), which was small enough for this large-scale shaking table test.

Provided that the rigid model container is significantly stronger than the HN31 sand model specimen, the side walls of the model container were then logically supposed to be infinitely rigid through setting the same X-displacement degree of

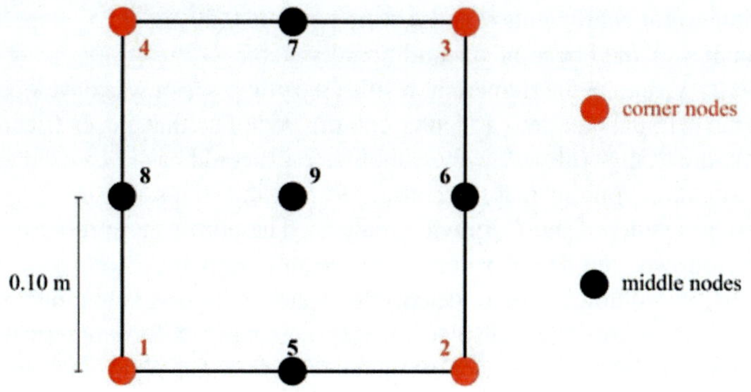

Fig. 5.43 Schematic of the nine four node Quad u-p element

freedom for all boundary nodes (see the solid red circles in Fig. 5.44). In the shaking table test, the top free surface is subjected to the total drainage condition: the associated degrees of freedom for fluid pressure of all the free surface nodes were blocked (no pressure changes). For all the base nodes, the vertical Y-displacements were blocked.

The simulation consisted of two distinct transient loading stages: (i) a first gravity analysis to initialize the soil sample with a natural consolidation state prior to the seismic excitation and (ii) a second dynamic analysis with possible sand liquefaction. The first stage was executed using huge time steps. Gravity was applied for the first 10 steps with $\Delta t = 2$ s to initialize the stress/strain states as a function of the consolidation stress. Once done, the elastoplastic behaviour of HN31 sand was considered using the Opensees command "updateMaterialStage" (https://opensees.berkeley.edu/wiki/index.php/UpdateMaterialStage), and the gravity analysis was repeated for another 100 steps with $\Delta t = 0.05$ so as to reduce the numerical divergence. The standard Newmark integrator was applied for the gravity analysis.

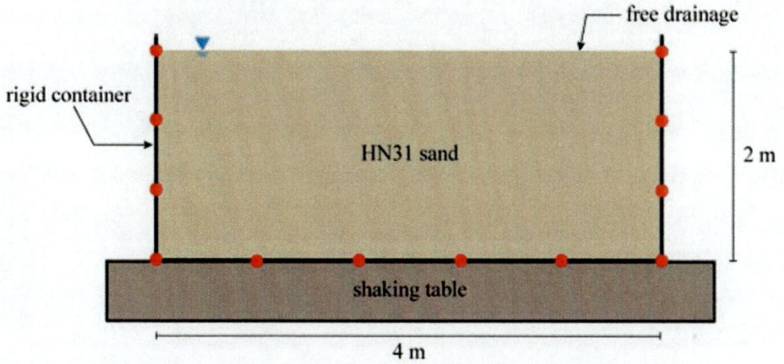

Fig. 5.44 Schematic of boundary conditions in the numerical simulation

The two integrator coefficients α and β were equal to 0.50 and 0.25, respectively so that the analysis would become unconditionally stable.

In order to visualize the numerical results, a Python script was developed which reshaped the original data into a regular column vector so that a user-friendly interactive software "GID" (https://www.gidhome.com/) could easily be used to display the corresponding contour maps according to the time steps. Figure 5.45 presents the numerical results of the *1*st gravity analysis. The pore water pressure shown in Fig. 5.45a followed the linear pattern of hydrostatic pressure $P = \rho_w \times g \times h$, in which ρ_w is the volumetric mass density for water, g is the gravitational acceleration and h is the height. The Y-displacement shown in Fig. 5.45b suggests that under gravity, the vertical displacement of the model specimen was very minor. In addition, a uniform distribution of both degrees of freedom was obtained for a given depth across the model specimen, indicating that the initial field for the latter dynamic simulation was successfully prepared.

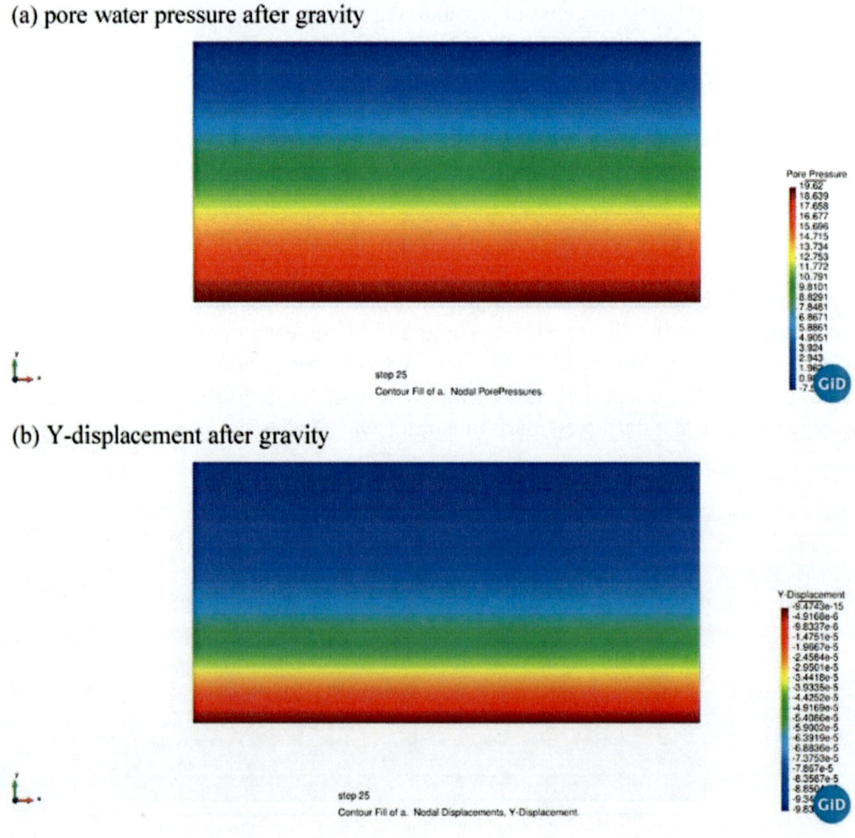

Fig. 5.45 Numerical results of model sand specimen after gravity analysis

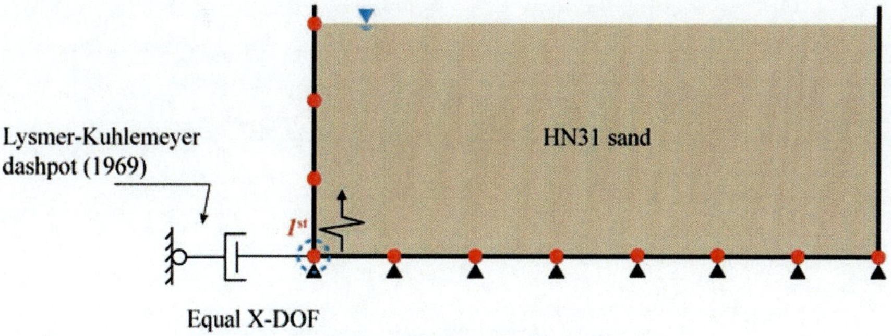

Fig. 5.46 Schematic of the Lysmer dashpot coupling with the *1*st node

For the dynamic analysis, a Lysmer dashpot (Lysmer and Kuhlemeyer 1969) was defined as a boundary condition, by using a "zeroLength" element connected to the first base node of the model sand specimen (coupling X-displacement degree of freedom), as displayed in Fig. 5.46. The horizontal seismic loading was applied as a force time history, following Joyner and Chen (1975) method, to the same node at the base. The force time history was converted by multiplying the velocity time history of an input motion by (i)–(ii) the volumetric mass density and the shear wave velocity of the "bedrock" (namely, the shaking table in the present work) and (iii) the base area of the model sand specimen. The clear advantage of this technique consists of considering the true rigidity of the "bedrock", instead of supposing it to be infinitely rigid without any additional energy dissipation. Similar to the gravity analysis, the dynamic analysis was executed as a transient calculation with the Hilber-Hughes-Taylor (HHT) integrator with $\alpha = 0.90$. Besides, a very small time step of $\Delta t = 0.10$ ms was applied to improve the accuracy. In the case of divergence, the time step Δt could be automatically reduced by a constant factor (e.g., a value of 2 used in this section) till the convergence criterion could be satisfied. Besides, a tolerance value $\Delta t_{\mathrm{tol}} = 5\mathrm{e}^{-7}$ was set as the minimum possible adaptive time step.

Figure 5.47 presents the comparison between the input motion (Peak Ground Acceleration PGA $= 2$ m/s^2) and the actual generated loading at the *1*st node: these curves were found almost identical with only minor differences in both the time and frequency domains. This fact indicates that the introduction of the Lysmer dashpot, following Joyner and Chen (1975) method, to apply the seismic loading as a force time history correctly worked.

Figure 5.48 shows the coloured contour fill of excess pore water pressure ratio r_{u} at $t = 8.0159$ s. It can be seen that the HN31 sand at the left top corner within the rigid model container was almost liquefied, attested by the r_{u} value of about 0.86 in the legend. With the use of the rigid container, the sand response was far from being uniform at a specific depth across the model specimen. And this phenomenon became prevailing while approaching the rigid side walls of the model container. The above observations might possibly be attributed to the generation of P-wave over the

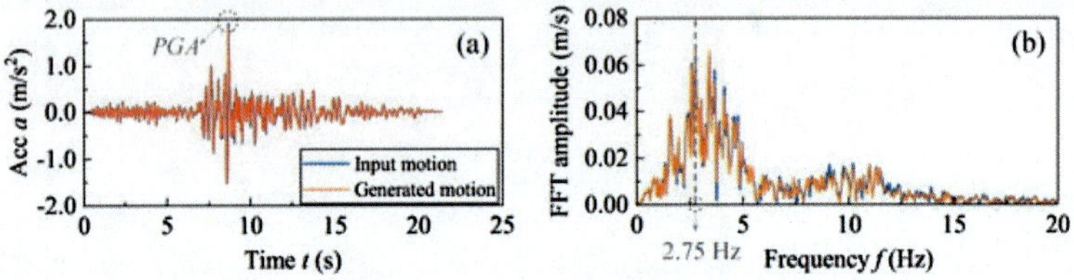

Fig. 5.47 Comparison between the input motion and generated motion at the 1st node in **a** time domain; **b** frequency domain

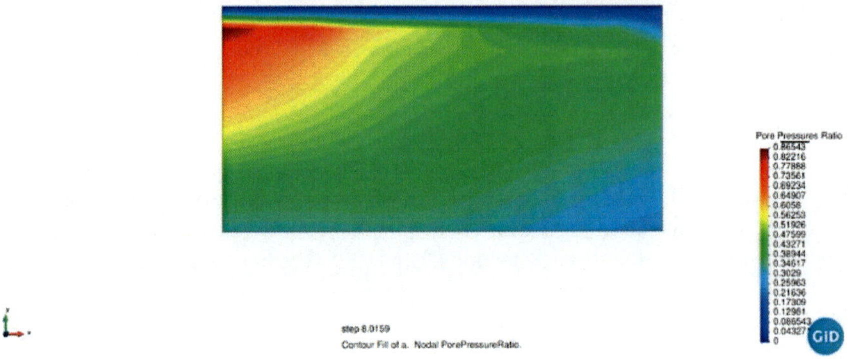

Fig. 5.48 Contour of excess pore water pressure ratio r_u during base shaking at $t = 8.0159$ s

rigid side walls during the base shaking. Thus, it is necessary to clearly examine the possible boundary effect and its possible influence areas.

Figure 5.49 presents the time histories of acceleration (after FFT bandpass filter with $f_{\text{cut1}} = 0.10$ Hz and $f_{\text{cut2}} = 15$ Hz) at three specimen profiles (side wall, mid and central parts) and at different depths. Near the free surface, the recording at the mid part was irregular with the presence of spikes, totally different from that observed at the left side wall and the central part. With the increment in *D*, the acceleration gap became gradually smaller, and the two recordings at the bottom could be viewed as almost the same. The above phenomena suggest that the boundary effect (including the generation of superfluous P-waves) is much prevailing near the free surface. However, the behaviour of granular sand near the bottom during shaking was much less polluted by the boundary effect, consistent with the real experiment reported by Li and Escoffier (2020), as shown in Fig. 5.7.

5.3 Development of the Large-sized Flexible Soil Container

As proved by the above numerical simulation and the relevant literature review, the model container composed of rigid side walls alone normally leads to a significant

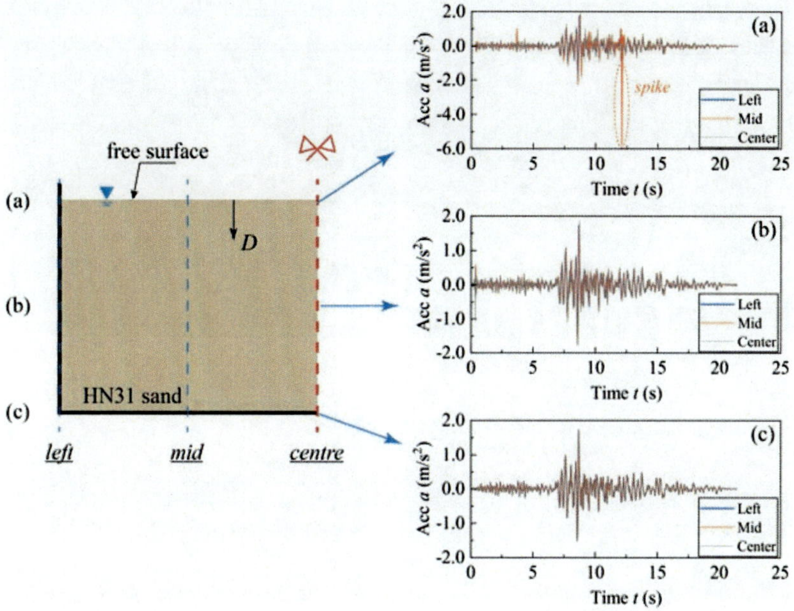

Fig. 5.49 Boundary effect on the acceleration readings at: **a** top free surface; **b** half-height; **c** bottom part

boundary effect, especially for the shaking table test involving the sand liquefaction phenomenon (Fig. 5.50).

For the purpose of restraining the superfluous wave reflections from the side walls to the model sand specimen during seismic base shaking, flexible foams of about 0.20 m were first mounted in the shaking direction on both sides. Later, a customized waterproof geomembrane (thickness of about 5 mm) was installed in order to prevent water from entering the porous foams. With this experimental configuration and such a large-sized model container, although the occurrence of wave reflections cannot be completely avoided, the central part of the sand specimen can be logically supposed to be less polluted and undergo a true simple shear condition subjected to the seismic base shaking.

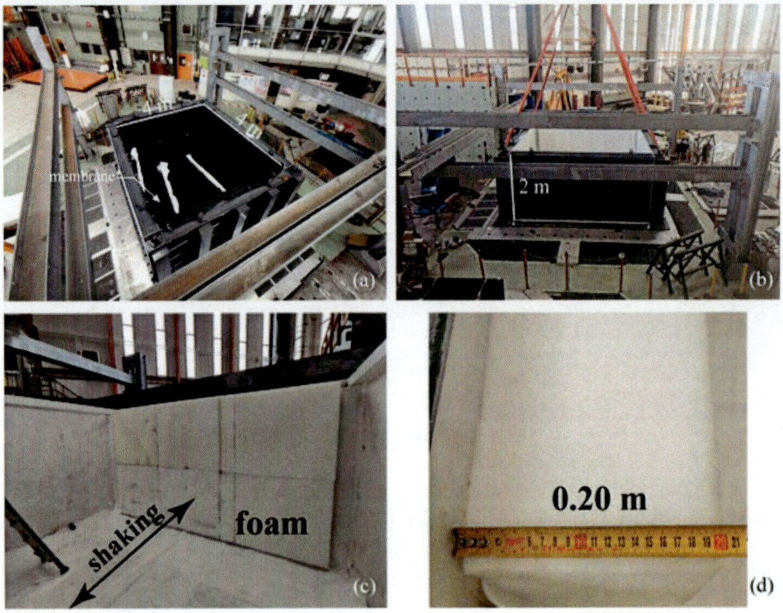

Fig. 5.50 Photographic views of **a–b** large-sized model container; **c–d** flexible foams of 0.20 m mounted on both sides

References

Armaghani DJ, Koopialipoor M, Marto A, Yagiz S (2019) Application of several optimization techniques for estimating TBM advance rate in granitic rocks. J Rock Mech Geotech Eng 11(4):779–789. Elsevier Ltd. https://doi.org/10.1016/j.jrmge.2019.01.002

Beaty MH, Byrne PM (2011) UbCSAND constitutive model

Been K, Jefferies M (1985) A state parameter for sands. Geotechnique 35(2):99–112

Benahmed N (2001) Comportement mécanique d'un sable sous cisaillement monotone et cyclique: application aux phénomènes de liquéfaction et de mobilité cyclique. Ph.D Thesis Ecole Nationale Des Ponts et Chaussées

Boulanger RW, Ziotopoulou K (2015) PM4SAND (Version 3): a sand plasticity model for earthquake engineering applications

Bratton D, Blackwell T (2008) A simplified recombinant PSO. J Artif Evol Appl 2008(1):1–10. https://doi.org/10.1155/2008/654184

Chang CS, Hicher PY (2005) An elasto-plastic model for granular materials with microstructural consideration. Int J Solids Struct 42(14):4258–4277. https://doi.org/10.1016/j.ijsolstr.2004.09.021

Chen C, Ueng T, Lee W (2004) Large scale biaxial shear box tests on shaking table. 13 th world conference on earthquake engineering, 1778, p 1778

Coelho PALF, Haigh SK, Madabhushi SPG (2003) Boundary effects in dynamic centrifuge modelling of liquefaction in sand deposits. In: 16th ASCE engineering mechanics conference. Seattle, pp 1–12

Dafalias YF (1986) Bounding surface plasticity. I: mathematical foundation and hypoplasticity. J Eng Mech ASCE, 112(9):966–987

Dafalias YF, Manzari MT (2004) Simple plasticity sand model accounting for fabric change effects. J Eng Mech 130(6):622–634. https://doi.org/10.1061/(asce)0733-9399(2004)130:6(622)
Eberhart RC, Shi Y (2000) Comparing inertia weights and constriction factors in particle swarm optimization. In: Proceedings of the 2000 congress on evolutionary computation. CEC, pp 84–88
Elgamal A, Yang Z, Parra E (2002) Computational modeling of cyclic mobility and post-liquefaction site response. Soil Dyn Earthq Eng 22(4):259–271. https://doi.org/10.1016/S0267-7261(02)00022-2
Howell R, Rathje EM, Boulanger RW (2015) Evaluation of simulation models of lateral spread sites treated with prefabricated vertical drains. J Geotech Geoenviron Eng 141(1):04014076. https://doi.org/10.1061/(asce)gt.1943-5606.0001185
Joyner W, Chen A (1975) Calculation of nonlinear ground response in earthquakes. Bull Seismol Soc Am 65(5):1315–1336
Kennedy J, Eberhart R (1995) Particle swarm optimization. In: Proceedings of ICNN'95—international conference on neural networks, pp 1942–1948
Kim H, Kim D, Lee Y, Kim H (2020) Effect of soil box boundary conditions on dynamic behavior of model soil in 1 g shaking table test. Appl Sci 10(4642)
Lee CJ, Wei YC, Kuo YC (2012) Boundary effects of a laminar container in centrifuge shaking table tests. Soil Dyn Earthq Eng 34(1):37–51. Elsevier. https://doi.org/10.1016/j.soildyn.2011.10.011
Lei H, Hu Y, Han Q, Zheng G, Zhao B, Du Y (2020) Design and test verification of a cylindrical 3D laminar shear soil container for use on shaking tables. Soil Dyn Earthq Eng 139(7):106384. Elsevier Ltd. https://doi.org/10.1016/j.soildyn.2020.106384
Li X, Dafalias Y (2000) Dilatancy for cohesionless soils. Geotechnique 50(4):449–460
Li XS, Dafalias YF, Wang ZL (1999) State-dependent dilatancy in critical-state constitutive modelling of sand. Can Geotech J 36(4):599–611. https://doi.org/10.1139/t99-029
Li XS, Wang Y (1998) Linear representation of steady-state line for sand. J Geotech Geoenviron Eng 124(12):1215–1217. https://doi.org/10.1061/(ASCE)1090-0241(1998)124:12(1215)
Li Z, Escoffier S (2020) Boundary effects in centrifuge modelling of soil liquefaction. 4th European conference on physical modelling in geotechnics. Luleå, Sweden, pp 39–42
Lysmer J, Kuhlemeyer A (1969) Finite dynamic model for infinite media. J Eng Mech 95:859–877
Manzari MT, Dafalias Y (1997) A critical state two-surface plasticity model for sands. Geotechnique 47(2):255–272
Mendez FJ, Pasculli A, Mendez MA, Sciarra N (2021) Calibration of a hypoplastic model using genetic algorithms. Acta Geotech 16(7):2031–2047. https://doi.org/10.1007/s11440-020-01135-z
Pal S, Wathugala GW, Kundu S (1996) Calibration of a constitutive model using genetic algorithms. Comput Geotech 19(4):325–348. https://doi.org/10.1016/S0266-352X(96)00006-7
Papadimitriou BAG, Bouckovalas GD, Dafalias YF (2001) Plasticity model for sand under small and large cyclic strains. J Geotech Geoenviron Eng 3(11):973–983
Rad NS, Tumay MT (1987) Factors affecting sand specimen preparation by raining. Geotech Test J 10(1):31–37
Ramirez J, Barrero AR, Chen L, Dashti S, Ghofrani A, Taiebat M, Arduino P (2018) Site response in a layered liquefiable deposit: evaluation of different numerical tools and methodologies with centrifuge experimental results. J Geotech Geoenviron Eng 144(10):04018073. https://doi.org/10.1061/(asce)gt.1943-5606.0001947
Richart F, Wood R, Hall J (1970) Vibration of soils and foundations. Int Ser Theor Appl Mech
Rowe P (1962) The stress-dilatancy relation for static equilibrim of an assembly of particles. In: Proceedings of the royal society of London. Ser A. Math Phys Sci 500–527
Shahir H, Pak A, Taiebat M, Jeremić B (2012) Evaluation of variation of permeability in liquefiable soil under earthquake loading. Comput Geotech 40:74–88. https://doi.org/10.1016/j.compgeo.2011.10.003
Shi Y, Eberhart RC (1998) Parameter selection in particle swarm optimization. In: Proceeding of evolutionary programming, pp 591–600

Suzuki H, Tokimatsu K, Sato M, Tabata K (2008) Soil-pile-structure interaction in liquefiable ground through multi-dimensional shaking table tests using e-defense facility. In: The 14 world conference on earthquake engineering

Tang L, Na SH (2021) Comparison of machine learning methods for ground settlement prediction with different tunneling datasets. J Rock Mech Geotech Eng 13(6):1274–1289. Elsevier Ltd. https://doi.org/10.1016/j.jrmge.2021.08.006

Trelea IC (2003) The particle swarm optimization algorithm: convergence analysis and parameter selection. Inf Process Lett 85(6):317–325. https://doi.org/10.1016/S0020-0190(02)00447-7

Turan A, Hinchberger SD, El Naggar H (2009) Design and commissioning of a laminar soil container for use on small shaking tables. Soil Dyn Earthq Eng 29(2):404–414. https://doi.org/10.1016/j.soildyn.2008.04.003

Ueng TS, Lee CA (2015) Pore pressure generation in saturated sand induced by one- and two-dimensional shakings. J GeoEng 10(2):53–61. https://doi.org/10.6310/jog.2015.10(2).3

Wang R, Zhang JM, Wang G (2014) A unified plasticity model for large post-liquefaction shear deformation of sand. Comput Geotech 59:54–66. Elsevier Ltd. https://doi.org/10.1016/j.compgeo.2014.02.008

Xie DY (2011) Soil dynamics. Higher Education Press, Beijing

Yang Z, Lu J, Elgamal A (2008) OpenSees soil models and solid-fluid fully coupled elements. University of California, San Diego

Yeh F, Chuang T, Ge L (2020) Calibration of advanced constitutive model using optimization techniques. J Test Eval 48(3):2196–2212

Yin ZY, Zhao JD, Hicher PY (2014) A micromechanics-based model for sand-silt mixtures. Int J Solids Struct 51:1350–1360

Zeng X, Schofield AN (1996) Design and performance of an equivalent-shear-beam container for earthquake centrifuge modelling. Geotechnique 46(1):83–102. https://doi.org/10.1680/geot.1996.46.1.83

Zhang JM, Wang G (2012) Large post-liquefaction deformation of sand, part I: physical mechanism, constitutive description and numerical algorithm. Acta Geotech 7(2):69–113. https://doi.org/10.1007/s11440-011-0150-7

Zhu Z, Cheng W (2020) Parameter evaluation of exponential-form critical state line of a state-dependent sand constitutive model. Appl Sci 10(1):328. https://doi.org/10.3390/app10010328

Zhu Z, Zhang F, Peng Q, Chabot B, Dupla J (2021b) Development of an auto compensation system in cyclic triaxial apparatus for liquefaction analysis. Soil Dyn Earthq Eng 144(December 2020):106707. Elsevier Ltd. https://doi.org/10.1016/j.soildyn.2021.106707

General Conclusions and Perspectives

At the end of this work, several conclusions and remarks can be formulated. For cyclic triaxial test involving loading reversals (from triaxial compression to extension and vice versa), the auto compensation system developed in the present work can largely avoid disturbing the fragile soil fabric prior to the shearing stage. Under the circumstances, a correct isotropic consolidation state can be achieved, which is always viewed as being beneficial for the subsequent liquefaction analysis. To demonstrate the robustness, several validation tests have been performed and a very good level of repeatability has been found. In this work, the density index of sand matrix was used as the controlling parameter to prepare the triaxial specimens containing fine particles, which is more consistent with field observations. The experimental results suggest that the mechanical role of added fines is highly related to the initial soil fabric associated with the selected reconstitution method. With the aid of microstructural observations, two schematic models have been proposed with the emphasis on different localisations of fine particles. This work highlights that for binary sandy deposits involving a predominant sand matrix and some small amounts of fine particles, the key point to assess its mechanical response is whether the presence of these fines can possibly break the sand-sand contacts. Furthermore, in the case of the dry tamping method, the weakening effect highly relies on the mean diameter ratio. This is because the finer added particles become, the more likely they can break the sand-sand contacts, leading to a significant decrease in sand liquefaction resistance. For simplicity, the impact of loading frequency and shape were studied with clean sand specimens. It was found that the inertia force is a kind of agency that needs to be seriously considered, especially approaching or already attaining the liquefaction stage with the development of transient displacement. As for loading shape, the rectangular loading caused the most severe liquefaction concern because of the greater loading acting for a longer duration. A novel parameter, derived from the Arias intensity, has been proposed to unify the sand liquefaction responses subjected to different loading shapes.

To get closer to in situ conditions involving a semi-infinite soil extent, two model soil containers have been developed for large-scale shaking table tests. With the first small one, a practical reconstitution and saturation method were planned and

Z. Zhu, *Influence of Fine Particles on the Liquefaction Properties of a Reference Sand*, Springer Theses, https://doi.org/10.1007/978-3-031-24299-1

carefully validated by the vane shear test and the measure of moisture content. In order to numerically assess the possible boundary effect during base shaking, a well-known bounding surface model was taken to represent the typical response of sand liquefaction. To better acquire the model parameters, an automatic calibration method based on artificial intelligence was developed in the Python environment. The versatility of the proposed programme has been validated by two reference sands. With the calibrated model parameters, a simulation of shaking table test has been performed. It was found that the boundary effect is a function of the depth. Namely, the occurrence of P-waves propagating in the horizontal direction from the rigid side walls to the model specimen becomes much more important while getting closer to the top free surface. With this in mind, a cost-effective solution was proposed. Several super flexible foams were then mounted on both the side walls in order to reduce the generation of P-wave during base shaking.

Due to the importance of the structural effects on the observed behaviours of sand-fines specimens, it should be undoubtedly essential to consider the scanning electron microscope (SEM) to keep a watchful eye on the possible localisation of fine particles especially during the shearing stage, which has been not covered in the present work due to the inherent limitation of the current experimental conditions. Besides, all fine particles selected in this work are much smaller than the host sand grain, creating quite large values of the mean diameter ratio. It might be very interesting to check what can take place when the size of added materials is close to the host sand grain.